Lecture Notes
in Business Information Processing **301**

Series Editors

Wil M.P. van der Aalst
Eindhoven Technical University, Eindhoven, The Netherlands
John Mylopoulos
University of Trento, Trento, Italy
Michael Rosemann
Queensland University of Technology, Brisbane, QLD, Australia
Michael J. Shaw
University of Illinois, Urbana-Champaign, IL, USA
Clemens Szyperski
Microsoft Research, Redmond, WA, USA

More information about this series at http://www.springer.com/series/7911

Ioannis M. Dokas · Narjès Bellamine-Ben Saoud
Julie Dugdale · Paloma Díaz (Eds.)

Information Systems for Crisis Response and Management in Mediterranean Countries

4th International Conference, ISCRAM-med 2017
Xanthi, Greece, October 18–20, 2017
Proceedings

 Springer

Editors
Ioannis M. Dokas
Department of Civil Engineering
Democritus University of Thrace
Xanthi
Greece

Narjès Bellamine-Ben Saoud (iD)
Ecole Nationale des Sciences de
    l'Informatique & Laboratoire RIADI
Université de la Manouba
Manouba
Tunisia

Julie Dugdale
Laboratoire d'Informatique
Université de Grenoble
Saint-Martin-d'Hères
France

Paloma Díaz
Departamento de Informática
Universidad Carlos III de Madrid
Madrid
Spain

ISSN 1865-1348   ISSN 1865-1356   (electronic)
Lecture Notes in Business Information Processing
ISBN 978-3-319-67632-6   ISBN 978-3-319-67633-3   (eBook)
DOI 10.1007/978-3-319-67633-3

Library of Congress Control Number: 2017953434

Printed on acid-free paper

This Springer imprint is published by Springer Nature
The registered company is Springer International Publishing AG
The registered company address is: Gewerbestrasse 11, 6330 Cham, Switzerland

Preface

A holistic and systematic analysis is typically required to control and respond effectively to a crisis, generated from man-made or natural causes. In such a context, the collaboration of agencies from different countries, as well as the exchange of data and information, is necessary. This is especially true for the Mediterranean counties that have historically strong geopolitical and cultural relations.

The ISCRAM-med conferences aim to enhance the collaboration and solidarity between Mediterranean countries in issues related to crisis management. They offer an outstanding opportunity to address and discuss new trends and challenges in the area of information systems and technologies for crisis response and management (ISCRAM). Special attention is given to innovative solutions and methods as well as to experiences from the design, development, and usage of information systems and technologies that support the management of the needs that emerge during the different phases of the crisis management lifecycle. The objective is to bring together researchers and practitioners from different countries to present their works and to discuss recent research results, their experiences, best practices, and case studies.

This volume contains the proceedings of the 4th ISCRAM-med Conference (ISCRAM-med 2017), which was held in the city of Xanthi, Greece, at the new facilities of the Civil Engineering Department of Democritus University of Thrace. Previous ISCRAM-med conferences were held in Toulouse, France (2014), in Tunis, Tunisia (2015), and in Madrid, Spain (2016).

ISCRAM-med 2017 attracted the attention of a significant number of researchers. The conference received 40 submissions involving 110 authors from 17 countries covering a broad area of topics. Out of the 110 researchers who authored one or more of the 40 submissions, 77 were from Mediterranean countries while 33 researchers were from non-Mediterranean countries.

More than 40 distinguished scientists and researchers from 17 countries formed the Program Committee members of the ISCRAM-med 2017 conference. The Program Committee members were responsible for the promotion of the conference and for the evaluation of the submitted manuscripts. Each submission was evaluated by the reviewers based on five criteria, namely, relevance, significance, originality, validity, and clarity. These are the same criteria on which the submitted manuscripts of the annual ISCRAM conferences are evaluated.

Based on the outcome of the evaluation process, the Organizing Committee decided to accept 12 full papers (12–14 pages long), which sets the acceptance rate of the conference to 30%. On top of these 12 papers, five short papers (up to 6-8 pages) were also accepted to be included in the proceedings and to be presented at the conference. When including short papers, the acceptance rate of the conference is 43%.

The 17 papers, which were selected to be presented during the 3 days of the conference, are organized in the following main topics: Robotic Systems for Crisis Management, Decision-Making in the Context of Crisis Management, Social

Networking and Big Data Analytics, Serious Games and Simulations, Collaboration and Information Sharing. Furthermore, the invited keynote speakers presented very interesting topics to the delegates of the conference ranging from the opportunities presented at the intersection of crisis management, sustainable development, and combinatorial optimization and risk analysis methods for crisis management, to real-life experiences from the ongoing refugee crisis.

We are grateful to the Steering Committee members and to all Program Committee members of the ISCRAM-med 2017 conference for their invaluable guidance and help and for their work on promoting and reviewing the submitted manuscripts.

Special thanks go to all members of the local Organizing Committee. We are grateful also to the Democritus University of Thrace for the financial support and for providing its new facilities as conference venue.

The local conference organizer acknowledges the help of Ms. Fisenchidou. Additionally, the local organizer would like to acknowledge the dean and the staff of the Polytechnic School and of the Civil Engineering Department of Democritus University of Thrace. Special thanks go to Profs. Iliadis, Gasteratos, Tsagarakis, and to Asst. Profs. Maris and Chalioris for their guidance and help.

Lastly, the Organizing Committee would like to thank the providers of the Easy-Chair conference system and everyone from Springer who helped us prepare this volume.

August 2017

Ioannis M. Dokas
Narjès Bellamine-Ben Saoud
Paloma Diaz
Julie Dugdale

Organization

Conference Co-chairs

Ioannis M. Dokas   Democritus University of Thrace, Greece
Chihab Hanachi   University of Toulouse 1, IRIT Laboratory, France
Paloma Díaz   Universidad Carlos III de Madrid, Spain
Julie Dugdale   Universitè Pierre Mendès France and Laboratoire
                            d'Informatique de Grenoble, France

Steering Committee

Chihab Hanachi   University of Toulouse 1, IRIT Laboratory, France
Frédérick Benaben   Ecole des Mines Albi Carmaux, France
François Charoy   University of Lorraine, France
Narjès Bellamine Ben   Ecole Nationale des Sciences de l'Informatique and
  Saoud   Laboratoire RIADI, Université de la Manouba, Tunisia
Julie Dugdale   Universitè Pierre Mendès France and Laboratoire
                            d'Informatique de Grenoble, France
Tina Comes   University of Agder, Norway
Victor Amadeo Banuls   Pablo de Olavide University, Spain
  Silvera

Program Co-chairs

Ioannis M. Dokas   Democritus University of Thrace, Greece
Narjès Bellamine Ben   Ecole Nationale des Sciences de l'Informatique and
  Saoud   Laboratoire RIADI, Université de la Manouba, Tunisia

Local Organizing Committee

Ioannis M. Dokas   Democritus University of Thrace, Greece
Fotios Maris   Democritus University of Thrace, Greece
Antonios Gasteratos   Democritus University of Thrace, Greece
Lazaros Iliadis   Democritus University of Thrace, Greece

Program Committee

Asim Abdulkhaleq   University of Stuttgart, Germany
Carole Adam   LIG CNRS UMR 5217 - UJF
Fred Amblard   IRIT, University of Toulouse 1 Capitole, France
Eric Andonoff   Irit/UT1
Ioannis Athanasiadis   Wageningen University, The Netherlands

Baghdad Atmani	Oran 1 University, Algeria
Elise Beck	Université Joseph Fourier, France
Narjes Bellamine	University of Manouba, ENSI, Tunisia
Lamjed Ben Said	ISG Tunis, Tunisia
François Charoy	Université de Lorraine, LORIA, Inria, France
Malika Charrad	Gabes University, Tunisia
Chantal Cherifi	Lyon 2 University, France
Jean Marie Dembele	Université Gaston Berger, Saint-Louis, Sénégal
Katerina Doka	National Technical University of Athens, Greece
Ioannis Dokas	Democritus University of Thrace, Greece
Julie Dugdale	Laboratoire d'Informatique de Grenoble, France
Paloma Díaz	Departamento de Informática - UC3M, Spain
Shady Elbassuoni	American University of Beirut, Lebanon
Mohammed Erradi	ENSIAS Rabat, Morocco
Daniela Fogli	Università di Brescia, Italy
Stephen Fortier	George Washington University, USA
Gaby Gurczik	German Aerospace Center, Germany
Chihab Hanachi	University of Toulouse 1, France
Muhammad Imran	Qatar Computing Research Institute, Qatar
Syed Imran	Furtwangen University, Greece
Demetrios Karras	Sterea Hellas Institute of Technology, Greece
Stefanos Katsavounis	Democritus University of Thrace, Greece
Fiona McNeill	Heriot Watt University, UK
Teresa Onorati	Universidad Carlos III de Madrid, Spain
Giorgos Papakostas	EMT Institute of Technology, Greece
Carmen Penadés	Polytechnic University of Valencia, Spain
Francois Pinet	Cemagref - Clermont Ferrand, France
Robert Power	CSIRO, Australia
Marco Romano	Universidad Carlos III de Madrid, Spain
Serge Stinckwich	IRD
Giorgos Sylaios	Democritus University of Thrace, Greece
Rui Jorge Tramontin	UDESC, Brazil
Erwan Tranvouez	LSIS, Polytech'Marseille Université d'Aix-Marseille, France
Giuliana Vitiello	University of Salerno, Italy
Stefan Wagner	University of Stuttgart, Germany
Telmo Zarraonandia	Universidad Carlos III de Madrid, Spain
Chris Zobel	Virginia Tech, USA

Additional Reviewers

Amri, Rahma	Kessentini, Maroua
Ben Yahia, Nesrine	Labba, Chahrazed
Hamami, Dalila	Laporte-Chabasse, Quentin
Hamed, Imen	Mindermann, Kai
Imran, Saed	

Abstracts of Invited Talks

Crisis Management and Sustainable Development Through Optimisation

Barry O'Sullivan

Insight Centre for Data Analytics, Department of Computer Science,
University College Cork, Ireland
barry.osullivan@insight-centre.org

Over the past decade the fields of big data analytics, artificial intelligence, and cloud-based computation, have changed the face of technological innovation in a variety of academic, societal, and industrial domains. In 2015 the United Nations agreed the Sustainable Development Goals (SDGs) focused on ending poverty, protecting the planet, and ensuring prosperity for citizens.[1] The challenges facing the world today can be categorised in many ways, but the following captures the general domains of concern:

- Health and well-being of citizens;
- Sustainable agriculture and forestry, food production, and marine;
- Energy management, generation, and renewables;
- Smart transportation systems;
- Environment and natural resource management;
- Sustainable manufacturing;
- Smart and inclusive government and policy-making;
- Data protection, rights to privacy, etc.

The role of data science, data analytics, and artificial intelligence, in making progress against these concerns is well accepted [9]. In this talk I will present an overview of the field of *computational sustainability* which focuses on the development of computational methods for balancing environmental, economic, and societal needs for a sustainable future [10]. Specifically, I will focus on the use of combinatorial optimisation [2, 11] in computational sustainability and, in particular, constraint programming, mixed integer programming, boolean satisfiability, quantified constraints, and quantified boolean formulae.

Crisis management plays a pivotal role in computational sustainability. The ubiquitous challenge is how to deal with disruptive unforeseen events [1]. I will present an overview of how combinatorial optimisation can be used to support crisis management considering modelling approaches and relevant solution concepts. For example, one can model complex problems using stochastic satisfiability [8], or quantified constraint satisfaction [4]. In stochastic settings one explicitly models uncertainty, while using

---

[1] http://www.un.org/sustainabledevelopment/sustainable-development-goals.

quantified approaches one distinguishes between variables that are under the control of the decision-maker and those that are set by nature.

In terms of solution concepts, there is a considerable literature on finding robust solutions to combinatorial problems [3]. One can look, for example, for fault tolerant solutions to a combinatorial problem. The concept here is to find a solution that can be repaired easily by changing some assignments. Such a solution is found in advance thereby pre-empting change or disruption. Therefore, I will focus on super models [5], super solutions [6], and weighted super solutions [7].

There is much to be gained at looking at the opportunities presented at the intersection of crisis management, sustainable development, and combinatorial optimisation. This talk will highlight some of these opportunities with the hope of motivating researchers to pursue this interesting multi-disciplinary area.

Acknowledgments. This work was supported by Science Foundation Ireland under Grant Number SFI/12/RC/2289.

References

1. Bénaben, F., Montarnal, A., Truptil, S., Lauras, M., Fertier, A., Salatgé, N., Rebiere, S.: A conceptual framework and a suite of tools to support crisis management. In: 50th Hawaii International Conference on System Sciences, HICSS 2017, Hilton Waikoloa Village, Hawaii, USA, 4–7 January 2017. AIS Electronic Library (AISeL) (2017). http://aisel.aisnet.org/hicss-50/cl/crisis_and_disaster_management/2
2. Biere, A., Heule, M., van Maaren, H., Walsh, T. (eds.): Handbook of Satisfiability, Frontiers in Artificial Intelligence and Applications, vol. 185. IOS Press (2009)
3. Brown, K.N., Miguel, I.: Uncertainty and change. In: Rossi et al. [11], pp. 731–760. https://doi.org/10.1016/S1574-6526(06)80025-8
4. Ferguson, A., O'Sullivan, B.: Quantified constraint satisfaction problems: from relaxations to explanations. In: Veloso, M.M. (ed.) Proceedings of the 20th International Joint Conference on Artificial Intelligence, IJCAI 2007, pp. 74–79, Hyderabad, India, 6–12 January 2007. http://ijcai.org/Proceedings/07/Papers/010.pdf
5. Ginsberg, M.L., Parkes, A.J., Roy, A.: Supermodels and robustness. In: Mostow, J., Rich, C. (eds.) Proceedings of the Fifteenth National Conference on Artificial Intelligence and Tenth Innovative Applications of Artificial Intelligence Conference, AAAI 1998, IAAI 1998, 26–30 July 1998, Madison, Wisconsin, USA, pp. 334–339. AAAI Press/The MIT Press (1998). http://www.aaai.org/Library/AAAI/1998/aaai98-047.php
6. Hebrard, E., Hnich, B., Walsh, T.: Super solutions in constraint programming. In: Régin, J.C., Rueher, M. (eds.) CPAIOR 2004. LNCS, vol. 3011. Springer, Heidelberg (2004). https://doi.org/10.1007/978-3-540-24664-0_11
7. Holland, A., O'Sullivan, B.: Weighted super solutions for constraint programs. In: Veloso, M.M., Kambhampati, S. (eds.) Proceedings of the Twentieth National Conference on Artificial Intelligence and the Seventeenth Innovative Applications of Artificial Intelligence Conference, pp. 378–383, 9–13 July 2005, Pittsburgh, Pennsylvania, USA. AAAI Press/The MIT Press (2005). http://www.aaai.org/Library/AAAI/2005/aaai05-060.php
8. Majercik, S.M.: Stochastic boolean satisfiability. In: Biere et al. [2], pp. 887–925. https://doi.org/10.3233/978-1-58603-929-5-887

9. Milano, M., O'Sullivan, B., Gavanelli, M.: Sustainable policy making: a strategic challenge for artificial intelligence. AI Mag. **35**(3), 22–35 (2014). http://www.aaai.org/ojs/index.php/aimagazine/article/view/2534

10. Milano, M., O'Sullivan, B., Sachenbacher, M.: Guest editors' introduction: special section on computational sustainability: where computer science meets sustainable development. IEEE Trans. Comput. **63**(1), 88–89 (2014). https://doi.org/10.1109/TC.2014.4

11. Rossi, F., van Beek, P., Walsh, T. (eds.): Handbook of Constraint Programming, Foundations of Artificial Intelligence, vol. 2. Elsevier (2006). http://www.sciencedirect.com/science/bookseries/15746526/2

A Systems Approach to Engineering Information Systems

John Thomas

Department of Aeronautics and Astronautics, Massachusetts
Institute of Technology, Cambridge, MA 02139, USA
jthomas4@mit.edu

KeyNote

As information systems become more advanced, it is becoming increasingly difficult to understand the complex organizational, technical, and human issues associated with their use. Traditional engineering approaches have not kept pace with the increasing complexity and the rapid introduction of advanced technologies, and we are seeing more and more interactions and behaviors during operation that were never foreseen during engineering and development. Problems like unexpected human-computer interaction issues, interoperability issues discovered only during operation, and inadequate or incomplete system requirements are becoming increasingly common.

A new system-theoretic approach has been developed to guide engineering and analysis of modern information systems. This qualitative model-based approach includes humans as a critical part of the overall system and is especially effective when applied during early development efforts to identify the necessary system requirements and drive critical design decisions. Although the proposed system-theoretic approach is very new to crisis response applications, it is already being used for safety-critical information systems in other industries including healthcare, aviation, automotive, space systems, chemical, and nuclear power.

Contents

Serious Games and Simulations

Collaboration and Information Sharing

Social Networking and Big Data Analytics

Comparative Study of Centrality Measures
on Social Networks

Nadia Ghazzali[⊠] and Alexandre Ouellet

Université du Québec à Trois-Rivières, Québec, Canada
{Nadia.Ghazzali,Alexandre.Ouellet}@uqtr.ca
http://www.uqtr.ca

Abstract. Many centrality measures have been proposed to quantify importance of nodes within their network [1,2]. This paper aims to compare fourteen of them: betweenness centrality, closeness centrality, communicability betweenness, cross clique centrality, in degree, out degree, diffusion degree, edge percolated component, eigenvector centrality, geodesic k-path, leverage centrality, lobby centrality, percolation centrality, semi-local centrality. Centralities are compared to their respective characteristics and with applications on two social networks. The first one is about communication within a terrorist cell [3], and the second concerns a sexually transmitted infection [4]. The main characteristics of each centrality measure have been identified. Centrality measures all succeed in identifying the most influential nodes on both networks. The results also show that measures slightly differ on non-predominant nodes.

Keywords: Statistical data analysis · Centrality measures · Social networks

1 Introduction

Social network analysis (SNA) have developed over the past decades as an integral part of advances in social theory, empirical research and formal mathematics and statistics [5]. The aim is to investigate the relationships among social entities, and on the patterns and implications between these relationships. More specifically, some research focused on determining the most influential entities of the social network by using centrality measures [6].

Jalili et al. [1] have identified 113 centrality measures of which 55 have implemented with a statistical free R package. We are interested in comparing fourteen of these 55 measures on two particular social networks. The first one represents a terrorist cell [3] of thirteen individuals. The second one is related to a sexually transmitted infection (STI) and concerns thirty-nine individuals [4]. Furthermore, the fourteen measures have different characteristics and computation times. Some of them have a time complexity too high to be used on large-scale network.

In this paper, we aim to compare the centrality measures in real case scenarios. We try to identify similarities between the value assigned by the centrality

© Springer International Publishing AG 2017
I.M. Dokas et al. (Eds.): ISCRAM-med 2017, LNBIP 301, pp. 3–16, 2017.
DOI: 10.1007/978-3-319-67633-3_1

measure to the individuals of the network. More specifically, we want to determine if the same most influential individual is identified by each of the fourteen centrality measures. Also, we want to quantify the dependencies between the ranks of the individuals in regards to the measure used.

The next section introduces graph concepts, and a summary of centrality measures used. We explain why we select the specific fourteen centrality measures. The Sect. 3 presents related works. The Sect. 4 contains the results and the analysis. A brief conclusion and references end this paper.

2   Graph and Centralities

We will modelize a social network by a graph, a usual representation on which can be applied centrality measures. Also, such graph representation offers an interesting visual support to understand the network structure.

2.1   Graph Definitions

A graph $G(V, E)$ is a set of nodes V, sometimes call vertices, and a set of edges E. If we want to refer to the nodes set of a graph G, we use the notation $V(G)$ (respectively $E(G)$ for the edges). N is the number of nodes also called the node set cardinality.

An edge is a pair of nodes (v_1, v_2). The first node (v_1) is called the origin node and the second (v_2) arrival node. A graph is directed if we can only go through an edge from the origin node to the arrival node. The nodes are graphically represented by dots, and the edges are represented by lines between the nodes. For directed graphs, edges are represented by an arrow from the origin node to the arrival node.

The degree of a node is the number of edges with the node as one of its extremities, noted $d(v)$. In a directed graph, we can define the entering degree and exiting degree. The entering degree (respectively exiting degree) of a node is the number of edges having this node as the arrival (respectively origin) node. We note the entering degree $d_{in}(v)$ and the exiting degree $d_{out}(v)$.

A path in a graph is an ordered edge set where any edge arrival node is the origin node of the following edge. The length of a path $\Delta(u, v)$ is the number of edges in it. We said that two nodes are adjacent (neighbours) if there is a path of length one between them. The function $\Gamma(v)$ gives the set neighbours of v, which is a subset of $V(G)$. A node is called peripheral if it only has one or no neighbour.

A geodesic path is the shortest path in the graph between two nodes. The function $\sigma_{u,v}$ counts the number of geodesic path between nodes u and v. We note the number of geodesic paths from node u to node v that go through the node r as $\sigma_{u,v}(r)$. The graph diameter is the longest of the geodesic paths in the graph. A sub-graph, $G'(V', E')$, is a graph for whom nodes set V' is a subset of V. Edges' subset, E', is composed of edges between nodes in V'. Not every edge between nodes of V' having to be in E'.

A connected component is a set of nodes for which there is an undirected path between each pair of nodes in the set. A directed graph has a strongly connected component if each pair of nodes in the component have a directed path between them. A clique in a graph is a sub-graph where every node is adjacent to any other node in the sub-graph (sometimes call complete sub-graph). Nodes within a certain radius r of another node is the set of nodes for whom exists a path of at most length r between them and the evaluated node.

Figure 1 presents an example of a graph. The nodes 1, 2, 3 and 4 are a clique of 4 nodes and nodes 6, 7 and 8 are a clique of 3 nodes. The graph has two connected components, one forms of nodes 9 and 10 and the other forms of other nodes. The bold path between nodes 3 and 8 is their geodesic path. Other paths exist between these nodes (e.g. 3–5–6–8), but the geodesic one is the shortest.

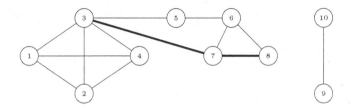

Fig. 1. An example of a graph.

2.2   Description of Centrality Measures

This section provides a brief summary of the centrality measures used in this research. A centrality measure is a function from $V \rightarrow \mathbb{R}$ that quantifies the importance of a node in its network. The highest values indicate most influential nodes.

For each considered centrality measures, a more complete reference is suggested. In fact, we propose four categories in order to compare each centrality measure. Table 1 shows the categories for each centrality measures. Thus, the measures in the category of geodesic path tend to assign high centrality values to the nodes involved in a large number of geodesic paths. For the distance category, the measures assign high values to nodes that are close (paths of small length) to the other nodes of the graph. The connectivity category considered the number of nodes' neighbours. Finally, the local influence category contains measures that consider only nodes in a certain distance of the evaluated node; at the opposite of the other centrality measures that consider the influence of all nodes in the graph or do not take account of the neighbours.

Betweenness Centrality (BC): This measure computes all geodesic paths between each pair of nodes and calculate the fraction of the shortest path going through the evaluated node. The extremity nodes are not considered involved

Table 1. Categories for centrality measures

Measure	Geodesic path	Distance	Connectivity	Local influence
Betweenness centrality	X			
Closeness centrality		X		
Communicability betweenness	X	X		
Crossclique centrality			X	X
Degree in			X	
Degree out			X	
Diffusion degree			X	X
Edge percolated component			X	
Eigenvector centrality		X	X	
Geodesic k-path	X	X		
Leverage centrality			X	X
Lobby centrality			X	X
Percolation centrality	X			
Semi-local centrality			X	X

in the geodesic path [7]. This measure evaluates the transmission of information in the graph based on the hypothesis that information only flow in the geodesic path.

We decide to include it because in some fields, such as communication networks, it is a reasonable to ensure that an actor who receives information transmits it quickly to the other, causes information to flow only through the shortest paths.

The centrality is defined for a given node $v \in V(G)$ as:

$$BC(v) = \sum_{v \neq r \neq s} \frac{\sigma_{r,s}(v)}{\sigma_{r,s}}, \; r, s \in V(G) \tag{1}$$

where $\sigma_{r,s}$ is the total number of geodesic paths from r to s and $\sigma_{r,s}(v)$ is the number of geodesic paths from r to s that having v as a node. Nodes r and s are different of v.

Closeness centrality (CL): The closeness between two nodes is the inverse of the path length between the two nodes. The distance is infinite if no path exists between the nodes [8].

We include this measure because it brings information on the position of the node in the graph. A node close to every other is visually central, and a node away from all others might be a peripheral node (or near to a peripheral node).

The closeness centrality for a node $v \in V(G)$ is given by

$$CL(v) = \sum_{u \neq v} \frac{1}{\Delta(u,v)}, \; u \in V(G). \tag{2}$$

where $\Delta(u,v)$ is the length of the path between nodes u and v.

Communicability Betweenness (CB): This measure is similar to the betweenness centrality, but also counts the other paths in the graph. The centrality consists of a weighted sum over all the possible paths within the graph. The longest paths have low rank assigned [9]. It can only be calculated for connected graph.

We considered this measure because it allows us not to make the assumption that information flows only through geodesic path, as it might be the case for networks such as the STI network. More information might be obtained in [9].

Cross Clique Centrality (CC): It counts for a node the number of cliques where it is involved in [10].

A node involved in a lot of cliques has many neighbours and large influence in sub-graphs. The value of the cross clique centrality is obtain by using an algorithm. The algorithm is given in [10].

Degree In and Out (DI, DO): It counts the number of edges with the node as the origin (arrival) node.

We include this measure because of its simplicity and it gives an information depending only on the evaluated node regardless of the graph organization.

Diffusion Degree (DD): It measures the capability of a node and of its immediate neighbours to transmit information in regards to a transmission probability λ [11].

We include this measure because in some application fields, such as infection propagation, the information is not always transmitted.

For a given node $v \in V(G)$ the measure is defined as

$$DD(v) = \lambda_v d(v) + \sum_{u \in \Gamma(v)} \lambda_u d(u) \tag{3}$$

where λ_k is the transmission probability of node k. The function $\Gamma(v)$ gives the set of the nodes neighbours to the node v.

Edge Percolated Component (EPC): To calculate this centrality, a proportion of edges are randomly removed from the graph. The value of the centrality is the average proportion of vertices that are still connected to the considered vertex [12].

This measure shows the impact of removing communication channel between individuals. Let us define E' the set of edges that remains in the graph after the removal step. The Kronecker delta function is defined on the set of initial edges as

$$\delta_{u,v} = \begin{cases} 0 & if \ (u,v) \notin E' \\ 1 & if \ (u,v) \in E' \end{cases} \tag{4}$$

The centrality measure is given for a particular node v

$$EPC(v) = \frac{1}{N} \sum_{u \in V(G)} <\delta_{u,v}> \tag{5}$$

where N is the number of nodes within the graph, and $<\delta_{u,v}>$ is the mean value of $\delta_{u,v}$.

Eigenvector Centrality (*EV*): This measure assigns each node a value proportional to its degree and to the measures' value of its neighbours. The centrality is influenced by the centrality of all the graph nodes. Nodes farther to the evaluated node have a lower influence on its centrality value. The vector obtained is proportional to the eigenvector of the greatest eigenvalue calculated on adjacency matrix [7].

This centrality measure gives information about how well a node can transmit information through the graph.

Geodesic k-Path (*GKP*): It counts the geodesic path of length at most k with the given node as the origin node [2]. Similar to betweenness centrality (see Eq. 1), but it allows to only use a certain radius around the evaluated node.

We decided to include it because it brings information on local diffusion of information. This characteristic may differ from the capability of transmitting information through the whole graph. The choose of the value k is important. For low value, only a small radius around the given node is considerate, so it does not show the influence of the graph organization. For high value of k, all nodes can reach all the other nodes of the graph, so that the values tend to be uniform and the centrality measure becomes irrelevant.

Leverage Centrality (*LC*): This measure counts the difference of the degree between a node and its neighbours. If in the average case, the node has a higher degree than its neighbours (positive centrality), the node has more influence on its neighbours than its neighbours have influence on it and vice versa [13].

This measure indicates which node is predominant in its own neighbourhood. We include this measure because it is the only one that gives information about the type of node behaviour (acting more like a leader or a follower).

For a given node $v \in V(G)$ the centrality is defined as

$$LC(v) = \frac{1}{d(v)} \sum_{u \in \Gamma(v)} \frac{d(v) - d(u)}{d(v) + d(u)} \tag{6}$$

where $d(k)$ is the degree of the node k.

Lobby Centrality (*LO*): This measure is the largest integer k such as the node has at least k neighbours of degree at least k [14]. It gives information about the graph configuration around the node. A high value indicates that the node is in a component with many influential nodes.

For a given node $v \in V(G)$ the lobby centrality is given by

$$\operatorname*{argmax}_{k \in \mathbb{N}} |\{v \in V(G)|d(v) \geq k\}| \geq k \tag{7}$$

where $d(v)$ is the degree of the node v.

Percolation Centrality (PC): It considers the diffusion of information through the graph [15,16] and might be evaluated a different step of the diffusion process. We choose to use the evaluated node as the first percolated node. At each step, previously percolated nodes percolates their immediate neighbourhood and then the centrality value is calculated.

The centrality calculation is similar to the betweenness centrality weighted by a factor indicating the percolation level of the graph. It also shares similarities with eigenvector centrality, but also considers the time process in the transmission of the information. This is an important measure because it is the only one that considers the time factor during the diffusion process.

The percolation state of a node indicates the node percolation level. We only consider where two percolation states: non-percolated (0) and percolated (1). None percolated nodes have not received the information and the percolated ones have received the information. The percolation state might be generalized for every number of states, even infinity.

The percolation centrality of node $v \in V(G)$ at time $t \in \mathbb{N}$ is given by

$$PC^t(v) = \sum_{r \neq s \neq v} \frac{\sigma_{r,s}(v)}{\sigma_{r,s}} \times \frac{x_s^t}{\sum_u x_u^t - x_v^t}, \quad u \in V(G) \tag{8}$$

where $\sigma_{r,s}$ counts the total number of geodesic paths between nodes r and s, $\sigma_{r,s}(v)$ counts the number of geodesic paths between nodes r and s going through the node v. The percolation state of the node v at time t is noted x_v^t.

Semi-local Centrality (SLC): This measure sums the number of neighbours to all immediate neighbours of the evaluated node [17]. It is similar to the diffusion degree but with a transmission probability of 1.

We included it because it gives a "worst case scenario" where the information is always spread through the graph.

The centrality for a given node $v \in V(G)$ is defined as

$$SLC(v) = d(v) + \sum_{u \in \Gamma(v)} Q(u) \tag{9}$$

$$Q(u) = \sum_{w \in \Gamma(u)} N(w) \tag{10}$$

where $d(v)$ is the degree of node v, $\Gamma(k)$ is the nodes set of node k neighbours and $N(w)$ is the number of nearest neighbours and next-nearest neighbours [17].

3   Related Works

Terrorist and STI networks had already been studied, [3,4,15,16] and some centrality measures had been calculated on these.

Azad et al. [3] studied the terrorist network. They evaluated four centrality measures: betweenness (BC), closeness (CL), degree (DI, DO) and eigenvectors

(EV). We reproduced their results for the specified four measures and extend the analysis to the ten other measures.

In addition, Hamed and Charrad [15] used only the percolation centrality on the terrorist network represented by a graph. They have compared the behaviour of peripheral nodes to non-peripheral nodes. We have reproduced part of their work but considered the directed graph.

De et al. [4] measured betweenness and degree centralities for the STI network. Piraveenan et al. [16] calculated the percolation centrality on a part of this network (the same part we study in the next section) for some particular initial conditions.

In this paper, we select fourteen centrality measures proposed in the literature [2,7,9,10,12–14,17]. Many of them seem to have been applied only to a particular field. Some centrality measures had been compared between them. It is the case of the betweenness (BC) and communicability betweenness (CB) [9], degree (DI, DO), closeness (CL), betweenness (BC) and cross clique (CC) [10], eigenvector (EV), leverage (LC), betweenness (BC) and degree (DI, DO) [13], degree (DI,DO), closeness (CL), betweenness (BC) and semi-local (SLC) [17], and closeness (CL), degree (DI, DO) and eigenvector (EV) [18]. At the best of our knowledge, no study takes account a large number of centrality measures and compare them on real world networks.

4   Application on Social Networks

For the calculation of the centrality measures, we use an R package [1] that proposes a large variety of them. Both networks are incomplete, because no member knows the whole network and some members of the network have interest in hiding information.

4.1   Networks Presentation

We calculate the centrality measures for two graphs. The first one is the terrorist network of the Mumbai attack (2006) [3]. This graph is directed. We choose to discard nodes 12 and 13 because, since they are not connected with the other nodes, they do not influence the value of the centrality. Furthermore, computing measures for the graph K_2, the complete graph with two nodes, are not relevant (Fig. 2).

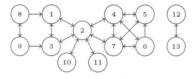

Fig. 2. Terrorist network

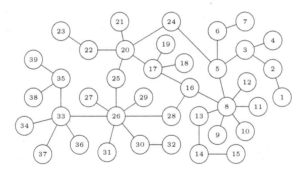

Fig. 3. STI network

The second graph is about the transmission of a particular sexual transmitted infection (STI) between thirty-nine individuals [4]. We analyzed the fourteen centrality measures. The graph is not directed because it is not always possible to know who was infected first and transmitted infection to the other. The graph study is the largest connected component extracted from a larger graph (the epidemic graph) (Fig. 3).

4.2   Evaluation of Critical Nodes

For the terrorist graph (see Table 2), we did not apply communicability betweenness (CB) and cross clique centrality (CC) because these measures require a strongly connected graph. The geodesic k-path (GKP) has been applied for $k = 2$, which is half of the graph diameter. The percolation centrality (PC) is calculated after the completion of two percolation steps. We choose this value because it is the maximum for the node 2 that percolates the graph in two steps. For the edge percolated component centrality (EPC), we used a removal probability of 0.5. The diffusion degree (DD) is calculated for a removal probability of 0.5. We calculated the values for other probabilities but the results are really similar and do not bring relevant information.

We calculated the rank of the nodes for each centrality measure. The nodes with the highest centrality value are assigned to the first rank. If there is a tie, the first rank possible is considered. The ranks of the terrorist network are presented in Table 2. We observed that node 2 has the first rank for every centrality measure and nodes 4 and 7 have the second-highest centrality value for almost every centrality measure. Nodes 1, 3, 5 and 6 have different ranks according to the measure used. This can be explained by the type of measure used and the graph topology. As an example, nodes 5 and 6 are in a dense connected component and nodes 1 and 3 cannot reach information to nodes 8 and 9. Leverage centrality (LC) shows that only nodes 2, 4 and 7 influence their neighbourhood which is consistent with the fact that they are the most influential nodes in the graph. We also observed that EPC is the only centrality measure that has no tie in its score.

Table 2. Centrality measures ranks for the terrorist network

Node	BC	CL	DI	DO	DD	EPC	EV	GKP	LC	LO	PC	SLC
1	4	4	4	5	6	6	6	2	6	6	2	4
2	1	1	1	1	1	1	1	1	1	1	1	1
3	4	4	4	5	6	7	6	2	6	6	2	4
4	2	2	2	2	2	2	2	2	2	1	2	2
5	8	6	4	5	4	4	4	8	8	4	8	6
6	8	6	7	4	4	5	4	8	8	4	8	6
7	2	2	2	2	2	3	2	2	2	1	2	2
8	6	8	8	5	10	11	10	8	4	6	6	10
9	6	8	8	5	10	10	10	8	4	6	6	10
10	8	8	8	10	8	8	8	6	10	10	8	8
11	8	8	8	10	8	9	8	6	10	10	8	8

For the STI network, we did not calculate out degree (DO) centrality because as it is a non-directed graph, its value is equal to the in-degree (DI). We calculated the geodesic k-path (GKP) with the value 5 which is half the graph diameter. Also, the percolation (PC) was calculated for 5 steps as the node 16 percolated the network in 5 steps. Node 16 is the one that percolated the STI network the fastest. The removal probability for the edge percolated component (EPC) is 0.5. The diffusion degree (DD) was calculated with a transmission probability of 0.5. We present the rank of each node in the Table 3.

Nodes 8 and 26 seem most important and have the rank of one or two for almost every centrality measures. Nodes 5, 16, 17, 20, 28 and 33 are also central and have high ranks for all measure, but not always in the same order. Less important nodes, such as every peripheral node, have lower rank. The lower ranks vary according to the centrality measure. The difference between the peripheral nodes is determined by their neighbour.

We observe that for both networks all centralities assign high value to the same nodes. The less important nodes have value varying according to the centrality measure.

4.3   Analysis of Centrality Measures

To compare the centrality measures between them, we use the Kendall correlation with the rank matrix presented in Tables 2 and 3. The values of Kendall correlation coefficients are presented in Table 4 for the terrorist network and in Table 5 for the STI network.

Given two paired ordered vectors of data $\mathbf{x} = \{x_1, x_2, \ldots, x_n\}$ and $\mathbf{y} = \{y_1, y_2, \ldots, y_n\}$, the Kendall correlation coefficient τ is defined as

$$\tau = \frac{2(P - Q)}{n(n - 1)}$$

Table 3. Centrality measures ranks for the STI network

Node	BC	CL	CB	CC	DI	DD	EPC	EV	GKP	LO	LB	PC	SLC
1	19	39	38	19	19	35	39	39	38	16	19	19	39
2	14	31	17	10	10	24	31	32	26	10	6	15	30
3	10	12	11	7	7	12	13	21	22	7	6	8	20
4	19	33	32	19	19	32	33	35	26	24	19	19	31
5	4	5	4	5	5	3	5	6	11	9	1	4	3
6	14	23	16	10	10	23	17	23	22	11	6	17	24
7	19	35	37	19	19	35	37	37	26	16	19	19	35
8	2	2	2	1	1	2	2	2	4	1	1	2	2
9	19	16	22	19	19	15	18	16	12	33	19	19	13
10	19	16	25	19	19	15	22	16	12	33	19	19	13
11	19	16	22	19	19	15	19	16	12	33	19	19	13
12	19	16	22	19	19	15	24	16	12	33	19	19	13
13	13	11	12	10	10	10	12	12	12	15	6	13	11
14	14	30	18	10	10	29	30	30	26	8	6	14	32
15	19	38	39	19	19	35	38	38	39	16	19	19	38
16	3	4	3	7	7	5	4	3	1	14	1	5	4
17	11	7	10	5	5	7	6	8	6	5	1	11	9
18	19	25	30	19	19	29	26	28	22	27	19	19	28
19	19	25	31	19	19	29	25	28	22	27	19	19	28
20	5	3	5	3	3	5	3	4	2	4	6	6	6
21	19	24	26	19	19	24	21	24	8	29	19	19	23
22	14	14	15	10	10	15	14	22	8	12	6	16	22
23	19	34	36	19	19	35	34	36	35	16	19	19	37
24	9	10	9	10	10	10	10	10	4	22	6	12	10
25	8	9	8	10	10	7	7	9	6	23	6	9	6
26	1	1	1	1	1	1	1	1	8	2	6	1	1
27	19	20	19	19	19	15	20	13	12	33	19	19	13
28	7	8	7	10	10	9	8	7	2	21	6	7	8
29	19	20	19	19	19	15	15	13	12	33	19	19	13
30	14	13	14	10	10	12	11	11	12	13	6	17	12
31	19	20	19	19	19	15	16	13	12	33	19	19	13
32	19	32	35	19	19	35	32	31	26	16	19	19	35
33	6	6	6	3	3	3	9	5	12	3	1	3	5
34	19	27	27	19	19	24	28	26	26	29	19	19	25
35	19	27	28	19	19	24	27	25	26	29	19	19	25
36	12	15	13	7	7	12	23	20	26	6	6	10	20
37	19	27	28	19	19	24	29	26	26	29	19	19	25
38	19	36	34	19	19	32	36	33	36	24	19	19	32
39	19	36	33	19	19	32	35	33	36	24	19	19	32

Table 4. Kendall's Tau for terrorism network

	BC	CL	DI	DO	DD	EPC	EV	GKP	LC	LO	PC	SLC
BC	1.00	0.74	0.70	0.67	0.46	0.44	0.46	0.68	0.79	0.53	0.96	0.63
CL	0.74	1.00	0.97	0.76	0.79	0.76	0.79	0.68	0.63	0.71	0.68	0.96
DI	0.70	0.97	1.00	0.75	0.84	0.82	0.84	0.64	0.59	0.76	0.64	0.93
DO	0.67	0.76	0.75	1.00	0.73	0.68	0.73	0.38	0.73	0.92	0.61	0.65
DD	0.46	0.79	0.84	0.73	1.00	0.95	1.00	0.57	0.36	0.77	0.39	0.84
EPC	0.44	0.76	0.82	0.68	0.95	1.00	0.95	0.54	0.34	0.73	0.37	0.80
EV	0.46	0.79	0.84	0.73	1.00	0.95	1.00	0.57	0.36	0.77	0.39	0.84
GKP	0.68	0.68	0.64	0.38	0.57	0.54	0.57	1.00	0.39	0.28	0.71	0.74
LC	0.79	0.63	0.59	0.73	0.36	0.34	0.36	0.39	1.00	0.60	0.74	0.52
LO	0.53	0.71	0.76	0.92	0.77	0.73	0.77	0.28	0.60	1.00	0.47	0.60
PC	0.96	0.68	0.64	0.61	0.39	0.37	0.39	0.71	0.74	0.47	1.00	0.57
SLC	0.63	0.96	0.93	0.65	0.84	0.80	0.84	0.74	0.52	0.60	0.57	1.00
% Correlated	75%	100%	100%	92%	75%	75%	75%	75%	66%	83%	66%	100%

Table 5. Kendall's Tau for STI network

	BC	CL	CB	CC	DI	DD	EPC	EV	GKP	LC	LB	PC	SLC
BC	1.00	0.72	0.84	0.91	0.91	0.73	0.67	0.63	0.52	0.63	0.88	0.96	0.64
CL	0.72	1.00	0.86	0.68	0.68	0.93	0.92	0.89	0.77	0.22	0.65	0.68	0.88
CB	0.84	0.86	1.00	0.77	0.77	0.88	0.81	0.78	0.65	0.28	0.75	0.79	0.81
CC	0.91	0.68	0.77	1.00	1.00	0.70	0.63	0.57	0.46	0.75	0.91	0.92	0.58
DI	0.91	0.68	0.77	1.00	1.00	0.70	0.63	0.57	0.46	0.75	0.91	0.92	0.58
DD	0.73	0.93	0.88	0.70	0.70	1.00	0.85	0.89	0.70	0.21	0.67	0.71	0.92
EPC	0.67	0.92	0.81	0.63	0.63	0.85	1.00	0.84	0.78	0.21	0.61	0.64	0.84
EV	0.63	0.89	0.78	0.57	0.57	0.89	0.84	1.00	0.77	0.12	0.55	0.59	0.94
GKP	0.52	0.77	0.65	0.46	0.46	0.70	0.78	0.77	1.00	0.09	0.49	0.48	0.75
LC	0.63	0.22	0.28	0.75	0.75	0.21	0.21	0.12	0.09	1.00	0.67	0.66	0.11
LB	0.88	0.65	0.75	0.91	0.91	0.67	0.61	0.55	0.49	0.67	1.00	0.87	0.56
PC	0.96	0.68	0.79	0.92	0.92	0.71	0.64	0.59	0.48	0.66	0.87	1.00	0.61
SLC	0.64	0.88	0.81	0.58	0.58	0.92	0.84	0.94	0.75	0.11	0.56	0.61	1.00
% Correlated	100%	92%	100%	100%	100%	92%	92%	92%	92%	54%	100%	100%	92%

where P is the number of observation pairs $(x_i, y_i), (x_j, y_j)$ where the ranks are concordant such as $x_i > y_i$ and $x_j > y_j$ or $x_i < y_i$ and $x_j < y_j$ (the ranks are in the same order for both observations). Q is the number of pairs where the ranks are not concordant. n is the number of observations [19].

Each coefficient had been submitted to the hypothesis test

$$H_0 : \tau = 0 \qquad H_1 : \tau \neq 0$$

We reject the null hypothesis if the p-value is under 5%. Bold values in Tables 4 and 5 are statistically significant.

We analyze centrality measures in regard to the categories previously defined at the Table 1. For the geodesic path category, the three measures have a significant correlation coefficient between them, which was expected. For the measures in the distance category, correlation coefficients of every pair are significant.

For the connectivity category, some correlation coefficients are not significant, especially for the leverage centrality (LC). This measure is not correlated with diffusion degree (DD), edge percolated component (EPC) and eigenvector centrality (EV). This might be explicated by the different information captured. LC uses only the node degree to determine the influence immediate neighbourhood. EPC and EV use the information from the whole graph. The main difference between DD and SLC is that DD assigns different weight to the node. The measures of the local influence category are all correlated between them except for LC and DD as discussed above.

The last line of Table 4 indicates the percentage of other measures that are significantly correlated with the evaluated measure. We notice that CL, DI and SLC are correlated with every other measures and LC and PC are the ones which have fewer significant correlation with other measures.

For the STI network almost every centrality measure is correlated to the other. The only centrality measure that is not significantly correlated is leverage centrality (LC). The findings are similar for the two networks.

Indeed, we observe similar tendencies in the behaviour of the measures for both networks, even though there are some differences between the networks. Centrality measures are for a vast majority correlated. These results are consistent with the interpretation of rank values of the Sect. 4.2. We can also affirm that there is also an association between the rank of many centrality measures even for none influential nodes.

5   Conclusion

We applied fourteen centrality measures on two social networks from different fields. A priori these two networks did not show similar characteristics but same behaviour from centrality measures has been observed. We can conclude that influential nodes are identified regardless of the centrality measure used. For non-predominant nodes, the value differs according to the type of centrality used and the graph organization. We also notice that ranks of the measures are highly correlated. Part of this correlation is explained by the strong link between influential nodes.

To the best of our knowledge, this work is the first to take account of a large number of centrality measures in order to compare them. Our results only applied to those two networks, and more studies are required to generalize them. Also, we need experts (sociologists, ...) of this kind of data to better interpret the results.

In future work we would deepen the percolation centrality at different steps and, therefore, we will study how information is spread in the network through time dynamically. Also, the percolation is currently calculated with a transmission probability of 1. We would analyze other transmission probabilities and compare them. Also, in order to generalize the results obtained in this paper, we plan to make a theoretical comparison of the measures and to replicate this study on larger networks.

References

1. Jalili, M., Salehzadeh-Yazdi, A., Asgari, Y., Arab, S.S., Yaghmaie, M., Ghavamzadeh, A., Alimoghaddam, K.: CentiServer: a comprehensive resource, web-based application and R package for centrality analysis. PloS one **10**(11), e0143111 (2015)
2. Borgatti, S.P., Everett, M.G.: A graph-theoretic perspective on centrality. Soc. Netw. **28**(4), 466–484 (2006)
3. Azad, S., Gupta, A.: A quantitative assessment on 26/11 Mumbai attack using social network analysis. J. Terrorism Res. **2**(2), 4–14 (2011)
4. De, P., Singh, A.E., Wong, T., Yacoub, W., Jolly, A.M.: Sexual network analysis of a gonorrhoea outbreak. Sex. Transm. Infect. **80**(4), 280–285 (2004)
5. Wasserman, S., Faust, K.: Social Network Analysis: Methods and Applications. Cambridge University Press, New York (1994)
6. Otte, E., Rousseau, R.: Social network analysis: a powerful strategy, also for the information sciences. J. Inf. Sci. **28**(6), 441–453 (2002)
7. Newman, M.: Network: An Introduction, p. 784. OUP, Oxford (2009)
8. Latora, V., Marchiori, M.: Efficient behavior of small-world networks. Phys. Rev. Lett. **87**(19), 198701 (2001)
9. Estrada, E., Higham, D.J., Hatano, N.: Communicability betweenness in complex networks. Physica A Stat. Mech. Appl. **388**(5), 764–774 (2009)
10. Faghani, M.R., Nguyen, U.T.: A study of XSS worm propagation and detection mechanisms in online social networks. IEEE Trans. Inf. Forensics Secur. **8**(11), 1815–1826 (2013)
11. Kundu, S., Murthy, C.A., Pal, S.K.: A new centrality measure for influence maximization in social networks. In: Kuznetsov, S.O., Mandal, D.P., Kundu, M.K., Pal, S.K. (eds.) PReMI 2011. LNCS, vol. 6744, pp. 242–247. Springer, Heidelberg (2011). doi:10.1007/978-3-642-21786-9_40
12. Chin, C.S., Samanta, M.P.: Global snapshot of a protein interaction network - a percolation based approach. Bioinformatics **19**(18), 2413–2419 (2003)
13. Joyce, K.E., Laurienti, P.J., Burdette, J.H., Hayasaka, S.: A new measure of centrality for brain networks. PLoS One **5**(8), e12200 (2010)
14. Korn, A., Schubert, A., Telcs, A.: Lobby index in networks. Physica A Stat. Mech. Appl. **388**(11), 2221–2226 (2009)
15. Hamed, I., Charrad, M.: Recognizing information spreaders in terrorist networks: 26/11 attack case study. In: Bellamine Ben Saoud, N., Adam, C., Hanachi, C. (eds.) ISCRAM-med 2015. LNBIP, vol. 233, pp. 27–38. Springer, Cham (2015). doi:10.1007/978-3-319-24399-3_3
16. Piraveenan, M., Prokopenko, M., Hossain, L.: Percolation centrality: quantifying graph-theoretic impact of nodes during percolation in networks. PloS one **8**(1), e53095 (2013)
17. Chen, D., Linyuan, L., Shang, M.S., Zhang, Y.C., Zhou, T.: Identifying influential nodes in complex networks. Physica A Stat. Mech. Appl. **391**(4), 1777–1787 (2012)
18. Koschtzki, D., Schreiber, F.: Comparison of centralities for biological networks. In: German Conference on Bioinformatics, pp. 199–206 (2004)
19. Kendall, M.G., Gibbons, J.D.: Rank Correlation Methods, p. 260. Edward Arnold, London (1990)

A Framework to Support Tunisian Tweets Analysis for Crisis Management

Jihene Sassi[1(✉)], Ines Thabet[1,2], and Khaled Ghedira[2,3]

[1] High School of Digital Economy, Campus Universitaire de la Manouba,
2010 Manouba, Tunisia
jihene.sassibarbouch@gmail.com, Ines.Thabetk@gmail.com
[2] Complex Outstanding Systems Modeling Optimization and Supervision,
COSMOS-SOIE, National School of Computer Sciences,
Campus Universitaire de la Manouba, 2010 Manouba, Tunisia
khaled.ghedira@anpr.tn
[3] Université Centrale, 3 Rue Hammadi Eljaziri, 1002 Tunis, Tunisia

Abstract. The increasing crisis frequency and the growing impact of their damages require efficient crisis management processes in order to manage crisis effectively and reduce losses. In such context, the need of accurate and updated information about crises is extremely important. In recent years, crisis information has frequently been provided by social media platforms such as Twitter, Facebook, Flickr, etc. In fact, considering the huge amount of shared information, their precision and their real time characteristic, organizations are moving towards the development of crisis management applications that include information provided by social media platforms. Following this view, the main purpose of our work is to propose a framework for Tunisian tweets extraction and analysis. More precisely, we provide an architecture that includes necessary components and tools for Tunisian dialect treatment. The proposed architecture is an extension on the existing AIDR platform. In addition, we specify the functioning of the proposed architecture to enable firstly terms transliteration from Arabic to Latin alphabet, secondly their normalization and finally their translation in order to be treated by existing social media analysis platform.

Keywords: Tweet analysis · Tunisian dialect · Crisis management

1 Introduction

The incidence and intensity of crises are increasing every day. In fact, both natural and man-made disasters are becoming a recurrent phenomenon and may be extremely damaging. They have destructive effects on both humanity and nature since they cause many detrimental consequences such as massive destruction of properties, high rates of death and have as a result substantial social, environmental and economic effects especially in developing countries.

In this context, a constant adaptation and optimization of crisis management processes is extremely important in order to respond to a crisis, save people's lives, preserve the environment and reduce crisis damages.

© Springer International Publishing AG 2017
I.M. Dokas et al. (Eds.): ISCRAM-med 2017, LNBIP 301, pp. 17–27, 2017.
DOI: 10.1007/978-3-319-67633-3_2

Crisis management is the process by which an organization deals with a major unpredictable event that threatens an organization or the whole society [1].

Accessing accurate, reliable and updated information is of primary importance for effective crisis management. During crises, information is vital for both civilian and emergency actors (police, firefighter, civil protection, etc.). When disaster strikes, people need to be informed about the situation in order to be out of the crisis, keep themselves safe and help others. For emergency actors accessing information is also of fundamental importance to face the crisis, build appropriate emergency plans and manage the crisis efficiently.

Nowadays, social media platforms such as Twitter, Facebook, Flickr and so on have become a reliable and useful source of information for crisis management. In fact, micro blogging platforms are actually considered as a means of emergency communication due to their growing ubiquity, rapidity of communication, and cross-platform accessibility [2].

In this context, Twitter, as one of the most popular micro blogging services, has recently received much attention. This online social network has been used to spread news about casualties and damages, donation offers, requests, alerts and so on, including multimedia information such as videos and photos [3–5]. It provides rapid access to situation-sensitive information that people post during crises (reports about disasters or accidents, videos, etc.).

Taking into account the huge amount of shared information, their precision and their real-time characteristics, emergency responders, humanitarian officials and other concerned organizations recognize the value of information posted on social media platforms during crises and are more and more interested in finding ways to incorporate it into their crisis management processes and procedures.

Given these reasons, crisis management applications based on tweet integration have recently been proposed. We can quote as examples "EMERSE: Enhanced Messaging for the Emergency Response Sector" [6] and "ESA: Emergency Situation Awareness Platform" [3] for tweet analysis and classification and therefore better disaster information relief, "Twitcident" for real-time tweets searching and current situation awareness [7], "AIDR: Artificial Intelligence for Disaster Response" [8] for automatic classification of crises-related micro blog communications, and so on. However, all these applications and platforms are designed primarily to process tweets written in English, French, Japanese, Chinese, etc. and none of these platforms may take into consideration tweets written in Tunisian dialect.

Since the last years, Tunisian dialect is widely used in new written media and web 2.0, especially in social network, blogs, forums, weblogs, etc., in addition to conversational media [9]. But, like most of dialects in the world, Tunisian dialect is considered as spoken language with no conventional written form. This dialect is very influenced by the French language. Tunisians' writing mode includes abbreviations, grammar and spelling errors, Arabic words written with French alphabets, numbers and several languages within a single tweet.

Given this context, the objective of our work is to propose a framework for Tunisian tweets analysis. We propose firstly an architecture that integrates necessary components and tools for Tunisian Dialect interpretation. The proposed architecture is an extension of the AIDR platform which is widely used in many operational deployments and has been mentioned by the United Nation Office for the Coordination of Humanitarian Affairs (OCHA) as one of the top 10 technologies relevant for the future of humanitarian response [10]. Secondly, we specify the functioning of our architecture for tweets collection, terms transliteration and normalization, and finally for normalized words translation.

The remainder of this paper is organized as follow. In the first section, we review related work. We firstly remind the general process for tweet extraction, analysis and visualization and secondly introduce some existing social media analysis frameworks for crisis management. In the second section, we give an overview of the proposed architecture and the components that we integrate in the AIDR platform to allow Tunisian dialect interpretation. We also specify accurately their functioning. Finally, we summarize and lay out the future work.

2 State of the Art

Social media is a general term that refers to data shared by its users such as information, photos and videos. Recently, social media like Facebook, Twitter, Flickr, and Myspace have attracted millions of people that integrated their services into their daily practices [11]. In fact, social networking sites are used for different purposes like staying in touch with friends and family, sharing interests and activities, connecting with others, expressing opinions, circulating relevant information expeditiously, etc.

Twitter, as one of the most popular micro blogging services, has recently received much attention. In fact, it was created in 2006, and was originally designed for mobile phones in order to allow users to send messages called tweets. However, given the huge number of Twitter subscribers, its high rate of usage and the fact that tweets are public, it became a key source of information for several recent studies. In fact, they contain a lot of potentially interesting information and their exploitation is one of the main axes of social network analysis. Their analysis perspectives may range from simple descriptive statistics (e.g. what are the most rewritten messages, the most followed users, etc.) to more sophisticated investigations (e.g. sentiment analysis, crisis management, event detection, etc.). Consequently, "Tweet analysis" has recently reached great importance and emerged as a new and important field of text mining. Researches include analyzing the site' network structure [12–14], studying Twitter for sentiment and emotion analysis [15, 16], evaluating social and cultural dimensions, creating new applications using Twitter and so on [17, 18].

In this section we firstly review the general process for tweet analysis and secondly give an overview of some existing platforms and applications for tweets analysis.

2.1 General Process for Tweet Analysis

The analysis of tweets is the process that goes from the *collection* of tweets, their *processing*, to their *visualization*. Tweet *collection* is an important task in the tweet analysis process. As a large social media platform, Twitter provides programmatic access to its content through Application Programming Interfaces (API). More specifically, three types of APIs are offered that are a Representational State Transfer (Rest) API, a Search API and a Streaming API. Their use is mostly guided by user's requirements. For example, the Rest API allows extracting the twenty most recent tweets sent on Twitter, information about its users and subscribers, its timeline, etc. The Search API allows querying data on Twitter based on keywords. Data can also be filtered by language, period of time and type of results. The Streaming API allows obtaining real-time tweets using a user specified query. Tweets can also be filtered by keywords, user identifier, geolocation and language. Generally, researches are mainly based on the Streaming API due to its real-time characteristic, its filtering criteria, and the important amount of data it provides. Once tweets are collected, several techniques may be used for tweets *processing* such as *classification, clustering* and *extraction*. In fact, tweets *classification* consists in classifying the tweets into a set of predefined classes using several algorithms such as Support Vector Machine, Naïve Bayes, K-nearest neighbors, etc. Tweets' *clustering* allows creating clusters so that elements within a cluster are more similar to each other than elements that belong to others clusters. Tweets' clustering is generally based on unsupervised machine learning algorithms such as K-means, hierarchical clustering, Density-Based Spatial Clustering of applications with Noise DBSCAN, etc. While information *extraction* allows extracting automatically structured information from unstructured or semi-structured documents such as plain text, web pages, etc. Finally, these processed data may be *visualized* via many tools to obtain maps, charts and boards.

2.2 Social Media Analysis Frameworks for Crisis Management

Twitter, as a tremendous source of real-time information, is currently one of the most used social media. It contains a lot of useful information that can be analyzed. As a result, many researchers have examined Twitter for crisis management, sentiment analysis, event detection, etc.

Concerning *crisis management*, several researches have dealt with particular crises. Takeshi Sakaki, Makoto Okazaki, and Yutaka Matsuo from the University of Tokyo proposed an algorithm to monitor tweets and detect a target event. They succeeded to develop an earthquake reporting system using tweets as event sensors. Tweets are firstly collected, then classified to detect earthquake and typhoon events and estimate their locations in Japan [18]. In [19], authors created a system that can report traffic information in real-time. They extracted traffic information from Twitter, classified them and broadcasted useful information to help people to plan their routes while avoiding traffic congestion in real-time.

Others researches have focused on the development of crisis management platforms. In addition to those mentioned in the introduction, other systems have been proposed such as *Twitcident* which is a Web-based system that automatically filters, searches and analyzes tweets regarding incidents [7]. Crisis Tracker [20] is a crowd-sourced social media platform for disaster awareness education. This tracker captures real-time distributed situation awareness reports based on social media activity during disasters [21]. It collects data from Twitter based on predefined filters (i.e., bounding box, keywords), then groups these tweets into clusters of tweets to be finally classified by humans. *Tweedr: "Twitter for Disaster Response"* is a Twitter-mining tool that extracts actionable information for disaster relief during natural disasters. The system classifies data to identify tweets regarding damages or casualties, then uses filters to merge tweets that are similar to others; and finally extracts tokens and phrases that report specific information about different classes of infrastructure damage, damage types, and casualties [19]. AIDR: "Artificial Intelligence for Disaster Response" is an online system for automatic classification of crises-related micro blog communications. It collects real-time tweets based on predefined filters (i.e., keywords, language, geolocation) and classifies them into a set of user-defined categories of information (e.g., \needs", \damage", etc.) [8].

Unfortunately, although all these systems seem very interesting, they treat crisis-related information in several languages such as English, French, and even Arabic but none of these systems can take in consideration our Tunisian Dialect. However, it is important to quote the work of Dridi et al. [22] for Tunisian dialect based tweets event detection. In their work, the authors propose an approach to deal with the specificities of our dialect. The proposed system allows extracting events that raised the interest of users within a given time period and the important dates for each event. Their approach is mainly based on three steps: data extraction from Twitter, terms grouping and finally event detection. During terms grouping a normalization is performed by their system using an adapted version of the Soundex algorithm. Their algorithm has been included in our work as mentioned in the Sect. 3 of our paper.

3   Overview of the Proposed Architecture for Tunisian Dialect Based Tweets Analysis

The area of crisis management in Tunisia is unfortunately still underdeveloped. Even though the majority of countries have specialized platforms in risk assessment and analysis, Tunisia doesn't have one. In this context, Twitter can be a useful and reliable source of information for both civilian and emergency actor. The huge amount of shared information can be effectively utilized for both risk assessment and analysis but also for preparedness and mitigation measures' optimization and crisis management improvement.

In this section, we introduce our architecture for Tunisian dialect based tweets analysis. Then, we describe its functioning. It is important to mention that we have adapted the widely used "AIDR" platform that allows the analysis and interpretation of disaster related tweets. More precisely we have proposed components that have been integrated in the AIDR platform. Each component is coping with a well-defined function and allows the pre-treatment of Tunisian dialect based tweets so they can be processed by the AIDR platform.

3.1   The AIDR Platform

The AIDR platform is free software for automatic classification of crises-related micro blog communications that can be run as a web application or downloaded to create its own instances. The AIDR platform functioning is based on the process described in the previous section. It allows collecting tweets in real time, and analyzing them based on keywords, language and geolocation. The AIDR platform allows collecting thousands of tweets and provides as a result an important amount of analyzed data. The AIDR platform classifies posted messages during disasters into a set of user-defined categories of information (e.g., \needs", \damage", etc.). More precisely, it collects crisis-related messages from Twitter (\tweets"), asks a crowd to label a subset of those messages, and performs an automatic classification based on defined labels [8]. The AIDR platform consists of three core components; a collector, a tagger, and a trainer. The *collector* is responsible for data collection using the Twitter streaming API described in the previous section. The *tagger* allows the classification of each individual tweet. Finally, the trainer manages the process of labelling items that are waiting to be labelled [23].

In the next section, we propose to integrate components and tools in the AIDR platform in order to perform a pre-treatment phase, so Tunisian based tweets can be analyzed by the platform.

3.2   The Proposed Architecture

Increasingly, Twitter is being used by millions of people around the world. This online social network limits the number of characters to 140 per tweet which encouraged users to use non-standard abbreviations and shorthand notations like 'srsly' to mean 'seriously' [24]. Generally, texts written by microblog users are often noisy and do not have formal settings, such us misspellings, lack of punctuation and grammatically incorrect sentences. Tunisian dialect, as a dialect used in social media, contains in addition to these specificities, Arabic words written in Latin, words containing numbers, and several languages in a single tweet. This variation in the Tunisian way of writing leads us to propose an extension of the AIDR platform to support Tunisian dialect. As shown in Fig. 1., our architecture includes the AIDR components that are the collector, the tagger and the trainer. Moreover, we integrate a *preprocessing module* responsible for a pretreatment phase. This module interacts with a *Tunisian knowledge base* and an *online translator*.

Let' us detail each component.

–   The ***transliterator***. Tunisian tweets are firstly gathered by the collector using the Twitter streaming API. It is important to notice that the collector proposed by the AIDR platform allows filtering tweets using keywords and/or hashtags. Thousands of tweets containing searched keywords and/hashtags are therefore collected in real time and according to specified keywords and hashtags, the collection of obtained data can reach up to hundreds of thousands of tweets. Obtained tweets are thereafter filtered to select only terms written in Arabic. In fact, the writing style of Tunisians is very variable. Tunisian tweets may be written in different ways. They may be expressed in Arabic or Latin letters, or by a mixture of text written with the Arabic alphabet and the Latin one. They may also include numbers that have specific letters as significations. While a specific algorithm is used by the normalizer to treat tweets including numbers, Arabic terms are transformed in Latin by the transliterator to be afterwards processed by the Normalizer. To do that, the transliterator applies a transliteration algorithm that implements a set of conversion rules that transform terms written in Non-Latin Script to their equivalent in Latin one. As a result, Arabic terms are transformed by the transliterator from Arabic script to Latin one and finally passed to the normalizer. Table 1 shows an example of Arabic letters and their equivalent Latin letters.

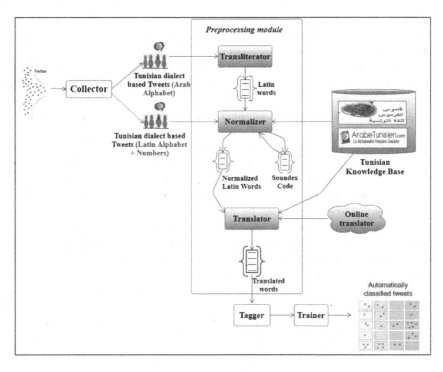

Fig. 1. Architecture for Tunisian dialect tweet based analysis

Table 1. Examples of Arabic letters and their corresponding Latin letters

Tunisian Word	Latin transcription
ب	b, p
ت	T
خ	Kh
و	W, ou
ف	F, v, ph

- The **normalizer**. In addition to the diversity in style of writing, Tunisians often commit misspellings and use abbreviations which create several varieties for the same term. These terms should however be interpreted as having the same meaning. For this purpose, as shown in Fig. 2., transliterated words from Arabic to Latin Alphabet, untreated collected tweets in Latin Alphabet and having Arabic signification (for example the word Harika is an Arabic word written in Latin Alphabet and meaning fire) and terms written by a mixture of Latin alphabet and number are processed by the normalizer. To do that, the normalizer uses a phonetic algorithm to index words by sound according to their pronunciation. More precisely the normalizer applies the Soundex algorithm [25] that converts selected word to a code. A same code is assigned to words written differently but having the same pronunciations to be interpreted by the platform as being the same keyword. For example, the terms e5wan and ikhwan will have the same code Soundex akhw [26]. Moreover, as we previously mentioned, the Tunisian dialect is very influenced by the French language which leads us to apply the Soundex algorithm for standard French. More specifically, we use a revised version of the Soundex proposed by Dridi et al. [27]. This algorithm has been adapted to support the Tunisian dialect and its specificities such as numbers included in words. In fact, even if the Soundex for standard French has succeeded in adequately normalizing terms, it has experienced difficulties in normalizing certain terms including the Romanized Arabic terms. Some modifications have therefore been applied to address these difficulties. More precisely, the revised version allows the replacement of numbers by their equivalent letters. For example, 9 is replaced by "k", 5 is replaced by "kh", 7 is replaced by "h", etc. Once Tweets are treated, the normalizer uses vocabulary from a proposed Tunisian dictionary to identify terms writing differently but having the same signification. The use of the dictionary allows also obtaining the meaning of treated words. To do that, the normalization algorithm is applied on both tweets and Tunisian words to obtain two sets of Soundex code: a set of code for collected tweets and another for Tunisian words. The objective is to obtain the same code for different words having the same pronunciation with their meaning in the dictionary.

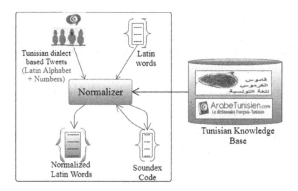

Fig. 2. The normalizer component

- The **Tunisian knowledge base.** As we previously mentioned, the normalizer allows detecting words writing differently but having the same meaning. Moreover, in order to have the definition of the Tunisian words produced by the normalizer, we have integrated two Tunisian dictionaries. This enables us to obtain a rich knowledge base and have more accurate and precise understanding of Tunisian words. The used dictionaries are the "karmous" dictionary and an online dictionary which is the "ArabeTunisien.com". The first one contains more than 3,800 Tunisian words, while the online one compromises more than 4,000 Tunisian word and expressions. A sample of index from Karmous Dictionary is introduced in Table 2. It is important to notice that only Tunisian words related to crises will be selected in our knowledge base since we are mainly interested in crisis management. Words written in Latin (French) are automatically collected by the AIDR platform.

Table 2. An index sample of Karmous dictionary

Tunisian Word	Latin transcription	Translation
تفجع	Tefjaa	être surpris et effrayé
شنوة	Chnowa	Quoi°
صبيطار	Sbitar	Hôpital

- The *translator.* As a last step for our pre-treatment phase, tweets are translated to French using the Tunisian Knowledge base. It is important to remind that the normalizer generates two sets of Soundex code one set for collected untreated tweets and one set of Tunisian Words from the knowledge base. These two sets of Soundex codes are firstly compared in order to identify similar codes and therefore similar words. Secondly, codes are retransformed again to their original words. Finally, they are translated to French using the Tunisian dictionaries from the knowledge base. Moreover, once normalized terms are translated in French, they can be translated again in any other language supported by the AIDR platform such as English, Japanese, etc. in order to be treated by the Trainer and the Tagger of the platform and finally classified. Online translators such as Google Translate, Bing Translator, Reverso.net and so on can be used for that purpose. Google Translate as one of the must used and simplest online translator can be suitable for our work. The tool allows detecting the word language automatically and offers instant translation. It also translates up to 30,000 signs at once and relies on hundreds of thousands of translations in its database to provide a suitable translation.

4   Conclusion and Future Work

"Crisis management is the application of strategies designed to help an organization to deal with a sudden and significant negative event" [28]. A crisis can occur as a result of an unpredictable event which requires that decisions are made quickly to limit damages. Dealing efficiency with a crisis requires not only defining optimized management plans but also obtaining accurate, reliable and real-time information about the crisis. Crisis management platforms are intended to realize such objectives. They can extract relevant information to detect early warning signals, identify incidents and threats, generate reports about a crisis or a foreseeable crisis, etc. This kind of information may help the emergency organizations to plan their emergency response and rescue in an efficient way. In recent years, several crisis management applications have been developed based on social media analysis, especially tweet analysis. Despite their high number and their diversity, none of these platforms can take into consideration the Tunisian dialect. Given this context and considering the critical importance of having such platform, we propose in our work a framework that supports our Tunisian dialect. More precisely, the proposed framework is an extension on the existing AIDR platform so Tunisian words can be treated by this platform. The proposed architecture enables firstly terms transliteration from Arabic to Latin alphabet, secondly their normalization and finally their translation in order to be treated by the AIDR platform.

This work opens a number of issues for future research. The first issue we are currently investigating is related to the ongoing implementation of the proposed framework in order to test its functioning. The second critical issue is related to the evaluation criteria of our work. Some measures such as the precision and recall metrics can be used to evaluate the efficiency of both transliterator and normalizer. These metrics are generally used to measure the system's effectiveness from the user perspective. Precision can be defined as the fraction of relevant instances among the retrieved instances, while recall is the fraction of relevant instances that have been retrieved over total relevant instances. The evaluation can also be realized by experts that can estimate the relevance of obtained results. Similar works [22, 29] have resorted the evaluation of their platform to experts in order to measure the accuracy of their translation. Finally, we intend to extend our framework in order to include a Tunisian map and be able to optimize the visualization of analyzed Tunisian Tweets.

References

1. Aldewereld, H., Tranier, J., Dignum, F., Dignum, V.: Agent-based crisis management. In: Guttmann, C., Dignum, F., Georgeff, M. (eds.) CARE 2009-2010. LNCS, vol. 6066, pp. 31–43. Springer, Heidelberg (2011). doi:10.1007/978-3-642-22427-0_3
2. Vieweg, S., Hughes, A.L., Starbird, K., Palen, L.: Microblogging during two natural hazards events what twitter may contribute to situational awareness. In: CHI 2010, Crisis Informatics, Atlanta, GA, USA (2010)
3. Cameron, M.A., Power. A., Robinson, B., Yin, J.: Emergency situation awareness from twitter for crisis management. In: Proceeding Conference on World Wide Web (WWW) (2012)
4. Imran, M., Elbassuoni, S., Castillo, C., Diaz, F., Meier, P.: Extracting information nuggets from disaster related messages in social media. In: Proceeding of ISCRAM, Baden-Baden, Germany (2013)
5. Qu, Y., Huang, C., Zhang, P., Zhang, J.: Microblogging after a major disaster in China: a case study of the 2010 Yushu Earthquake. In: Proceeding of CSCW (2010)
6. Caragea, C., McNeese, N., Jaiswal, A., Traylor, G., Kim, H.W., Mitra, P., Wu, D., Tapia, A., Giles, L., Jansen, B., Yen, J.: Classifying text messages for the Haiti earthquake. In: Proceedings of the 8th International Conference on Information Systems for Crisis Response and Management (ISCRAM 2011) (2011)
7. Abel, F., Hauff, C., Houben, G.J., Stronkman, R., Tao, K.: Semantics + filtering + search = twitcident. exploring information in social web streams. In: Proceedings of the 23rd ACM Conference on Hypertext and Social Media, Milwaukee, Wisconsin, USA, pp. 285–294 (2012)
8. Imran, M., Castillo, C., Lucas, J., Meier, P., Vieweg, S.: AIDR: artificial intelligence for disaster response. In: Proceedings of the 23rd International Conference on World Wide Web (WWW). Seoul, Korea, pp. 159–162 (2014)
9. Diab, M., Habash, N.: Arabic dialect processing tutorial. In: Proceedings of the Human Language Technology Conference of the NAACL, pp. 5–6 (2007)
10. LAMDA Learning And Mining from DatA. http://lamda.nju.edu.cn/seminar_150717.ashx?AspxAutoDetectCookieSupport=1
11. Boyd, D., Ellison, N.B.: Social network sites: definition, history, and scholarship. J. Comput.-Mediated Commun., 210–230 (2007)

12. Java, A., Song, X., Finin, T., Tseng, B.: Why we twitter: an analysis of a microblogging community. In: Zhang, H., Spiliopoulou, M., Mobasher, B., Giles, C.L., McCallum, A., Nasraoui, O., Srivastava, J., Yen, J. (eds.) SNAKDD/WebKDD -2007. LNCS, vol. 5439, pp. 118–138. Springer, Heidelberg (2009). doi:10.1007/978-3-642-00528-2_7

13. Huberman, B.A, Romero, D.M, Wu, F.: Social Networks that Matter: Twitter Under the Microscope. ArXiv E-Prints (2008)

14. Kwak, H., Lee, C., Park, H., Moon. S.: What is twitter, a social network or a news media?. In: Proceeding 19th International Conference World Wide Web (WWW 2010), pp. 591–600 (2010)

15. Boyd, D., Golder, S., Lotan, G.: Tweet, tweet, retweet: conversational aspects of retweeting on twitter. In: Proceeding of 43rd Hawaii International Conference System Sciences (HICSS-43) (2010)

16. Tumasjan, A., Sprenger, T.O, Sandner, P.G., Welpe, I.M.: Predicting elections with twitter: what 140 Characters Reveal About Political Sentiment. In: Proceeding of Fourth International AAAI Conference on Weblogs and Social Media (ICWSM) (2010)

17. Galagan, P.: Twitter as a Learning Tool. Really, ASTD Learning Circuits (2009)

18. Okazaki, M., Matsuo, Y., Sakaki, T.: Tweet analysis for real-time event detection and earthquake reporting system development. IEEE Trans. Knowl. Data Eng. **25**(4), 919–931 (2012)

19. Wanichayapong, N., Pruthipunyaskul, W., Pattara-Atikom, W., Chaovalit, P., Social-based traffic information extraction and classification. In: 11th International Conference on ITS Telecommunications, St. Petersburg, Russia (2011)

20. Rogstadius, J., Vukovic, M., Teixeira, C., Kostakos, V., Karapanos, E., Laredo, J.A.: CrisisTracker: crowdsourced social media curation for disaster awareness. IBM J. Res. Dev. **57**(5), 4:1–4:13 (2013)

21. Ashktorab, Z., Brown, C., Nandi, M., Culotta, A.: Tweedr: mining twitter to inform disaster response. In: Proceedings of the 11th International ISCRAM Conference, Pennsylvania, USA (2014)

22. Dridi, H.E., Lapalme, G.: Détection d'évènements à partir de Twitter. TAL. Réseaux Sociaux **54**(3), 11–33 (2013)

23. Imran, M., Castillo, C., Lucas, J., Meier, P., Rogstadius, J.: Coordinating Human and Machine Intelligence to Classify Microblog Communications in Crises. In the 11th International Conference on Information Systems for Crisis Response and Management Pennsylvania, USA, (2014)

24. Ahmad, B.: Lexical normalisation of twitter data. In: Science and Information Conference (SAI), London, UK (2015)

25. Russell, R.: Soundex coding system. United States Patent (1918)

26. Dridi, H.E.: Détection d'évènements à partir de Twitter, PhD thesis, Université de Montréal, Montréal, Canada (2014)

27. Imran, M., Castillo, C., Diaz, F., Vieweg, S.: Processing social media messages in mass emergency: a survey. J. ACM Comput. Surv. (CSUR) Surv. Homepage Arch. **47**(4) (2015). Article No. 67

28. WhatIs.com. http://whatis.techtarget.com/definition/crisis-management

29. Bouchlaghem, R., Elkhlifi, A.: Tunisian dialect Wordnet creation and enrichment using web resources and other Wordnets. In: Proceedings of the EMNLP 2014 Workshop on Arabic Natural Langauge Processing (ANLP), Doha, Qatar (2014)

Exploiting Social Networking and Mobile Data for Crisis Detection and Management

Katerina Doka[1]([envelope]), Ioannis Mytilinis[1], Ioannis Giannakopoulos[1],
Ioannis Konstantinou[1], Dimitrios Tsitsigkos[2], Manolis Terrovitis[2],
and Nectarios Koziris[1]

[1] CSLab, NTUA, Athens, Greece
{katerina,gmytil,ggian,ikons,nkoziris}@cslab.ece.ntua.gr
[2] IMIS, RC Athena, Athens, Greece
{tsitsigkosdim,mter}@imis.athena-innovation.gr

Abstract. Every day, vast amounts of social networking data is being produced and consumed at a constantly increasing rate. A user's digital footprint coming from social networks or mobile devices, such as comments, check-ins and GPS traces contains valuable information about her behavior under normal as well as emergency conditions. The collection and analysis of mobile and social networking data before, during and after a disaster opens new perspectives in areas such as real-time event detection, crisis management and personalization and provides valuable insights about the extent of the disaster, its impact on the affected population and the rate of disaster recovery. Traditional storage and processing systems are unable to cope with the size of the collected data and the complexity of the applied analysis, thus distributed approaches are usually employed. In this work, we propose an open-source distributed platform that can serve as a backend for applications and services related to crisis detection and management by combining spatio-textual user generated data. The system focuses on scalability and relies on a combination of state-of-the art Big Data frameworks. It currently supports the most popular social networks, being easily extensible to any social platform. The experimental evaluation of our prototype attests its performance and scalability even under heavy load, using different query types over various cluster sizes.

1 Introduction

In the recent years we have witnessed an unprecedented data explosion on the web. The wide adoption of social networks has concluded in terabytes of produced data every day. In March 2017 for example, Facebook had on average 1.28 billion daily active users [5], while more than 500 million tweets are produced on a daily basis [7]. The proliferation of mobile, smart devices has contributed decisively to this trend: With more than half the world now using a smartphone, active mobile social media users account for 34% the total population [3].

As this data is mostly a product of human communication, it reveals valuable information covering all aspects of life, even emergency situations. Indeed, social

© Springer International Publishing AG 2017
I.M. Dokas et al. (Eds.): ISCRAM-med 2017, LNBIP 301, pp. 28–40, 2017.
DOI: 10.1007/978-3-319-67633-3_3

media have become a prevalent information source and communication medium in cases of crises, producing high throughput data only seconds after a crisis occurs, as attested by the 500k tweets produced in the first hours after the tsunami in Philippines and the 20k tweets/day registered during the Sandy storm in New York in 2012 [16].

Thus, the processing and linkage of such mobile social media information, which includes heterogeneous data varying from text to GPS traces, provide tremendous opportunities for the detection of emergency situations [15], for the provision of services for immediate crisis management (e.g., getting life signs from people affected [4], communicating with responders, etc.) and for data analytics that can assist in short and long-term decision making by evaluating the extent of the disaster, its impact on the affected population and the rate of disaster recovery [10].

When the volume, velocity and variety of the collected data or the complexity of the applied methods increase, traditional storage and processing systems are unable to cope and distributed approaches are employed. To this end, we propose, design and implement a distributed, Big Data platform that can power crisis detection and management applications and services by combining heterogeneous data from various data sources, such as user GPS traces from cell phones, profile information and comments from existing friends in various social networks connected with the platform. Currently our prototype supports Facebook, Twitter and Foursquare, but it can be extended to more platforms with the appropriate plug-in implementation. It has been tested with real data but simulated workloads (i.e., synthetic user base).

Through distributed spatio-temporal and textual analysis, our system provides the following functionalities:

- Socially enhanced search of crisis-related information based on criteria such as location, time, sentiment or a combination of the above.
- Automatic discovery of new Points of Interest (*POIs*) and events that could indicate an emerging crisis of any extent, small or big, ranging from traffic jams and spontaneous gatherings such as protests, to natural disasters or terrorist attacks.
- Inference of the user's semantic trajectory during and after the emergency through the combination of her GPS traces with background information such as maps, check-ins, user comments, etc.
- Semi-Automatic extraction of a user's activity during the crisis in the form of a blog.

An important feature of our system is the automatic POIs detection. A distributed version [8] of DBSCAN, a well-known clustering algorithm, is applied to the GPS traces of our system's users. A dense concentration of traces signifies a POI existence. Furthermore, the correlation of spatio-temporal information provided by the GPS traces with POI related texts automatically produces a blog with the user's activity.

Moreover, a user can search for crisis-related social media information posing both simple as well as more advanced criteria. Simple criteria include commonly used features such as keywords (e.g., "flood", "terrorism", "traffic", etc.), location (e.g., a bounding box on a map) or a time frame of interest. Advanced criteria refer to data annotations, that derive from the processing of data, such as the sentiment of a tweet or a Facebook post. On top of these, each search can be socially charged, taking into account one's social graph when providing responses (e.g., rank content based on the sentiment of one's friends of it). Thus, the proposed platform is capable of supporting queries such as "Which are the places in Greece where protests are currently taking place and my Facebook friends (or a subset of them) participate in" or "Inform me of the activity and sentiment of my Facebook friends that were near Lesvos island on June 12, 2017 (when the earthquake of 6.1 Richter scale occurred)".

The contribution of this work is many-fold:

- We design a highly scalable architecture that efficiently handles data from heterogeneous sources and is able to deal with big data scenarios.
- We adapt and fine-tune well-known classification and clustering algorithms in a Hadoop-based environment.
- We experiment with datasets in the order of tens of GB, from Tripadvisor, Facebook, Foursquare and Twitter.
- We validate the efficiency, accuracy and scalability of the proposed architecture and algorithms.

The remainder of this paper is organized as follows: Sect. 2 presents the system architecture, Sect. 3 provides an experimental evaluation of the platform, Sect. 4 presents the related work and Sect. 5 concludes our work.

2   Architecture

Our platform follows a layered architecture that is illustrated in Fig. 1. It comprises the *frontend* and *backend* layers, both of which follow a completely modular design to favor flexibility and ease of maintenance.

The frontend layer constitutes the point of interaction with the user and includes all applications related to crisis detection and management that can be supported by the platform. Such an application can be a web, a mobile phone (e.g., Android, iOS) or a native application. To test the platform's basic functionality, a web application has been implemented.

The applications communicate with the backend layer through a REST API. The requests to the backend as well as the responses of the backend back to the user follow a specific JSON format. This feature enables the seamless integration of any client applications with the platform.

The backend is divided in two subsystems, the *processing subsystem* and *storage subsystem*. For the processing subsystem, a Hadoop cluster and a web server farm have been deployed, in order to cater for the special requirements of social network data processing.

Indeed, the volume and velocity of the data produced by social networks demand a distributed approach. Since the Hadoop framework has emerged as

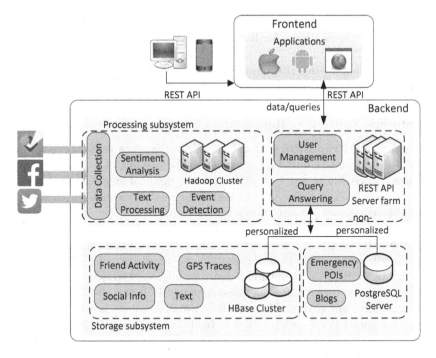

Fig. 1. The proposed platform architecture.

the most prevalent platform of choice for large-scale analytics, we design and deploy the following Hadoop-based processing modules: (a) the *Data Collection* module, (b) the *Sentiment Analysis* module, (c) the *Text Processing* module and (d) the *Event detection* module.

The web server farm hosts the *User Management* and the *Query Answering* modules, which act as gateways to the platform. Both modules are implemented as lightweight web services which put load to the datastore without stressing the web servers.

The storage subsystem is responsible for storing all the data used by our platform, both raw and processed. We refer to the components of the storage subsystem as *repositories*. Repositories are conceptually classified to *primitive* and *non-primitive* data repositories. Primitive repositories store raw, unprocessed data, referred to as *primitive data*. Primitive data are collected from external data sources such as social networks (Facebook, Foursquare, Twitter) and GPS traces and are directly stored to the platform. Non-primitive data repositories are the ones serving answers to queries and hold information extracted from the analysis of primitive data through the use of spatio-textual algorithms.

To handle multiple concurrent users that issue queries to the platform, the system follows a scalable approach. To this end, the Apache HBase [1] NoSQL datastore is used. However, there are queries which require either complex indexing schemes or extended random access to the underlying data. These queries cannot be efficiently executed in HBase. For this reason, we devise a hybrid architecture that uses HBase for batch queries that can be efficiently executed in parallel and PostgreSQL [6] for online random-access queries that cannot.

In the following, we describe in more detail the repositories and the processing modules of the proposed platform.

2.1 Storage Subsystem Repositories

Emergency POI repository: It is a non-primitive data repository that contains all the information our platform needs to know about POIs where an emergency has occurred. The name of a POI, its geographical location, the keywords characterizing the emergency and the sentiment-related metrics are all stored in this repository. A new entry to the repository can be inserted either explicitly by the user through the web GUI or automatically by the Event Detection module. While POI repository has to deal with low insert/update rates, it should be able to handle heavy, random access read loads. Thus, indexing capabilities are required. PostgreSQL offers such capabilities and has therefore been chosen as the ideal infrastructure for hosting the emergency POI repository.

Social Info repository: This primitive data repository is a HBase-resident table where social graph information is held. For each platform user and for each connected social network, the list of friends is persisted. More specifically, we store a compressed list with the unique social network id, the name and the profile picture of each friend. This list is periodically updated through the Data Collection module to capture possible changes in a user's social graph.

Text repository: The textual data collected from social media an processed through the Text Processing module are stored in this non-primitive data repository. Since the anticipated volume of this data type is high, the Text repository is most demanding in terms of disk space requirements. For this reason, it is stored in HBase and spread across all available cluster nodes. The Text repository holds all the collected comments and reviews that contain crisis related keywords, such as "flood", "hurricane", "traffic", "protest", etc. along with their geo-location. Texts are indexed by user, geo-location and time. For any given crisis-related keyword and any given rectangle on the map, we are able to retrieve the comments that any user made at any given time interval, containing the keywords and geo-located within the search area.

Friend Activity repository: In order to give information based on social friends' activity, we need to keep track of all locations (Emergency POIs) and social media content of a user's friends. This information is maintained in the Friend Activity repository, a non-primitive data repository persisted as an HBase table. Each activity is represented by an *activity* data structure with the complete Emergency POI information (name, latitude, longitude, etc.) as well as the textual content of possible posts. Moreover, this structure is enriched with sentiment metrics (positive/negative) through the Sentiment Analysis module. Every time a user or a user's social friend is tracked near an Emergency POI through her GPS trace or geo-tagged posts, an activity struct indexed by user and time is added to the repository. Thus, for any given time interval, we know the places that all of a user's friends have been and a score indicating each friend's sentiment.

An obvious remark is that the activity struct introduces high data redundancy, since each time someone visits an Emergency POI, the whole POI information is registered within the struct. The alternative schema design strategy would be joining POI information with activity information at query time. However, our experiments suggest data replication to be more efficient. Our schema in combination with HBase coprocessors and a fully parallel query mechanism offers higher scalability and achieves lower latency in the event of many concurrent users, which is the case in emergency situations. Thus, we sacrifice cheap storage space for efficiency.

GPS Traces repository: Mobile devices with appropriate geo-location applications supported by our platform installed, can push their GPS traces to the platform. These traces are stored in the primitive data repository of GPS Traces. Since the platform may continuously receive GPS traces, this repository is expected to deal with a high update rate. Furthermore, as GPS traces are not queried directly by the users but are periodically processed in bulk, there is no need to build indices on them. The volume of data, the opportunities for parallel bulk processing and the absence of indices are the main reasons why we choose HBase as the storage substrate for GPS data.

Blogs repository: We define a semantic trajectory to be a timestamped sequence of Emergency POIs summarizing user's activity during crisis this information is stored in a non-primitive data repository, the Blogs repository. As POIs, blogs are frequently queried by users but they do not have to deal with heavy updates and thus are stored as a PostgreSQL resident table.

2.2 Processing Subsystem Modules

User Management module: The User Management module is responsible for the user authentication to the platform. The user is registered either through the mobile applications or the website. The signing-in process is carried out only with the use of the social network credentials. The registration workflow follows the OAuth protocol. The OAuth authorization framework enables a third-party application to obtain access to an HTTP service on behalf of a resource owner. When the authentication is successful, the user logs in and an access token is returned to the platform. With this token, the platform can interact with the connected social networks on behalf of the end user. It can monitor user's activity, user's friends activity, posts etc. When multiple social networks are connected to the platform, the platform joins the acquired data and enriches the information that is indexed and stored.

Data Collection module: The functionality of this module is to collect data from external data sources. Periodically, the Data Collection module scans in parallel all the authorized users of the platform; each worker scans a different set of users. For each user and for all connected social networks, it downloads all the interesting updates from the user's social profile. Since the platform provides social geo-location services, interesting updates are considered to be user check-ins and the accompanying comments, status updates and geo-located tweets. From this information, the platform is able to gain knowledge about the existence of crisis events and people's sentiment about them. Once data are streamed to the platform, part of them are indexed and stored in the primitive data repositories, while the rest are processed in-memory and then indexed and stored to the appropriate non-primitive data repositories.

Text Processing module: This module indexes all textual information collected by the Data Collection module according to predefined crisis related keywords (e.g., flood, earthquake, protest, etc.). To do so, it employs standard Natural Language Processing (NLP) techniques (lemmatization, stemming, etc.) to track the predefined keywords and stores the results to the Text repository.

Sentiment Analysis module: The Sentiment Analysis module performs sentiment analysis to all textual information the platform collects through the Data Collection module. Comments are classified, real-time and in-memory, as positive or negative. The score which results from the sentiment analysis is persisted to the datastore along with the text itself.

As a classification algorithm, we choose the Naive Bayes classifier that the Apache Mahout [2] framework provides. Naive Bayes is a supervised learning algorithm and thus it needs a pre-annotated dataset for its training. For the training, data from Tripadvisor, containing reviews for POIs is used. The reason for choosing this training set is that Tripadvisor comments are annotated with a rank from 1 to 5 that can be used as a classification score. After an extensive experimental study and a fine-tuning of the algorithm parameters, we managed to create a highly accurate classifier that achieves an accuracy ratio of 94% towards unseen data.

Event Detection module: The detection of new events and emergency POIs constitutes a core functionality of our platform. A distributed, Hadoop-based implementation of the DBSCAN clustering algorithm [8] is employed for this reason. The module is called periodically and processes in parallel the updates of the GPS Traces repository in order to find traces of high density; high density traces imply the existence of a new POI. In order to avoid detecting already known Emergency POIs, traces falling near to existing Emergency POIs in the repository are filtered out and are not taken into consideration for clustering.

Query Answering module: The Query Answering is the module that executes search queries. A search query can take as input the following parameters:

- a bounding box of coordinates (i.e., on a map)
- a list of keywords
- a list of social network friends
- a time window
- results sorting criteria
- the number of results to be returned

Queries are distinguished in *personalized* and *non-personalized* ones. Personalized are the queries that exclusively concern the social media friends (or a subset of them) of a user. Thus, if a list of friends is provided, the query is considered to be *personalized*.

For personalized queries, the selected friends' activities should be taken into account. The repository that maintains such personalized information is the Friend Activity repository which resides in HBase. Thus, as Fig. 1 depicts, personalized queries are directed towards the HBase cluster. In order to efficiently resolve personalized queries, HBase coprocessors are used. Each coprocessor is responsible for a region of the Friend Activity repository and performs HBase *get* requests to the users under its authority. Since different friends are located with high probability in different regions, a different coprocessor is in charge of serving their activities and multiple get requests are issued in parallel. Increasing the number of regions leads to an increase in the number of coprocessors and thus a higher degree of parallelism is achieved within a single query.

On the other hand, non-personalized queries involve repositories that handle no-personalized information. The Emergency POI repository, which contains all the required information, resides in PostgreSQL. Thus all non-personalized queries are translated to *select* SQL queries and directed towards PostgreSQL.

3   Experiments

We experimentally evaluate our system in terms of performance, scalability and accuracy of its modules. This way we validate both our architectural design and the selected optimizations.

3.1   Scalability and Performance Evaluation

We first present some experiments for the scalability and performance of the query answering module. Using a synthetic dataset, we test the ability of the platform to respond to personalized queries issued by its users for various loads and different cluster sizes. Each personalized query involves a set of friends and returns the activities, i.e., text, geo-location and sentiment related to a crisis situation designated by a keyword (e.g., earthquake). The platform user can set a number of other parameters as well: the geographical location, the time period of the activities, etc. In our experiments, we identified that the dominant factor in the execution time of the query is the number of social network friends that the platform users define.

For the synthetic dataset generation, we collected information from Open-StreetMap about 8500 POIs (which we considered Emergency POIs for out experiment) located in Greece. Based on those POIs, we emulated the activity of 150k different social network users, each of whom has performed activities in a number of Emergency POIs and assigned a sentiment to it. The number of activities for each social network friend follows the Normal Distribution with $\mu = 170$ and $\sigma = 10$[1]. The dataset is deployed into an HBase cluster consisting

---

[1] The vast majority of the users has performed between 140 and 200 visits in different POIs.

of 16 dual-core VMs with 2 GB of RAM each, running Linux (Ubuntu 14.04).
The VMs are hosted in a private Openstack cluster.

At first, we are going to study the impact of the number of social network
friends into the execution time of a single query. At this point we will also
examine how the cluster size affects the execution time of a query. Secondarily,
we are going to extend our study to multiple concurrent queries where we will
also examine the behavior of the platform for different numbers of concurrent
queries and different cluster configurations.

For the first point, we evaluate the execution time of a query for different
numbers of friends. In Fig. 2 we provide our findings. In this experiment, we
executed one query at a time involving from 500 to 10 k social network friends
for three different cluster setups consisting of 4, 8 and 16 nodes. The friends for
each query are picked randomly in a uniform manner. We repeated each query
ten times and we provide the average of those runs.

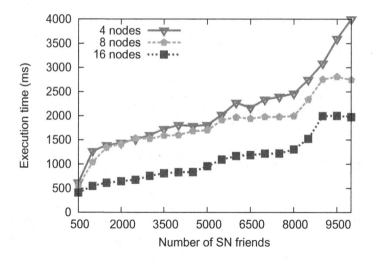

Fig. 2. Query latency vs number of users

The number of friends affects the execution time in an almost linear manner.
Furthermore, an increase in the cluster size leads to a latency decrease, since
the execution is happening in parallel to multiple nodes. By utilizing HBase
coprocessors, we managed to exploit the locality of the computations into specific
portions of the data: each coprocessor operates into a specific HBase region
(holding a specific portion of the data), eliminates the activities that do not
satisfy the user defined criteria, aggregates multiple activities referring to the
same Emergency POI and annotates them with aggregated sentiment scores.
Finally, each coprocessor returns to the Web Server the list of emergency POIs,
the related activities and sentiment and the Web Server, in turn, merges the
results and returns the final list to the end user.

Using the previously described technique, we achieved latencies lower than 1 second for more than 5000 users. Bearing into consideration that social networks like Facebook, retain a limit on the maximum number of connections (5000 friends per user), we can guarantee that the latency for each query remains acceptable for a real time application.

We now extend our analysis for the cases where multiple queries are issued concurrently to the platform. For our experiments we create a number of concurrent queries involving 6000 social network friends each and we measure their execution time for different cluster sizes. In Fig. 3, we provide our results. The execution time in the vertical axis represents the average execution time for each case.

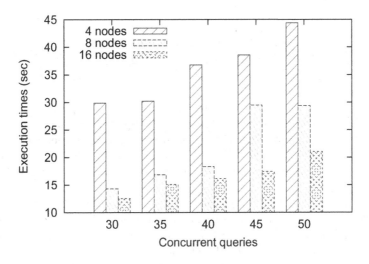

Fig. 3. Average execution time for concurrent queries

As Fig. 3 demonstrates, an increase in the number of concurrent queries leads to worse performance (larger execution time). However, for larger cluster sizes we can make the following observations: (a) even for the lowest number of concurrent queries the 16 cluster case is approximately 2.5 times better than the 4 cluster case, indicative of the proper utilization of more resources and (b) larger cluster sizes do not allow execution time to rise fast as the number of concurrent queries increases. Specifically, when the cluster consists of 4 nodes, the execution time is high even for the lowest number of concurrent queries and it continues to rise rapidly while the number of queries is increased. In the 8 nodes case, although at first the achieved execution times are relatively low, the increase becomes rapid for more concurrent queries, whereas in the 16 nodes case we see that the increase is held in a minimum level. This is indicative of the scalability of the platform, since more resources are properly utilized and the platform becomes resistant to concurrency.

Finally, since greater number of concurrent queries leads to more threads in the Web Server which, in turn, hits the cluster, we can avoid any potential bottlenecks by replicating the Web Servers while simultaneously, we use a load balancer to route the traffic to the web servers accordingly. In our experimental setup, we identified that two 4-cores web servers with 4 GB of RAM each are more than enough to avoid such bottlenecks.

3.2 Evaluation of Sentiment Analysis

In this section we evaluate the tuning of the Naive Bayes classifier we use for sentiment analysis. As we have already mentioned, for the training of the classifier, we crawl and use data from Tripadvisor. We consider a Tripadvisor review about a place to be a classification document. We divide training documents into two sets: positive and negative opinion documents. Both sets should have almost the same cardinality. Before feed the training set to the classifier, a preprocessing step is applied which involves stemming, turning all letters to lowercase and removing all words belonging to a list of stopwords. When the preprocessing step is finished, Naive Bayes is applied to the data. Let this procedure be the baseline training process. As a next step, we experiment with the following optimizations: use of the *tf* metric, *2-grams*, *Bi-Normal Separation* and deletion of words with less than x occurrences. These optimizations can be given as parameters to the classification algorithm. Experiments with different combinations for the algorithm parameters were also conducted but are omitted due to space constraints.

Figure 4 shows the classification accuracy for various training set sizes, when the baseline and the optimized classification are used. We observe that when optimizations are applied, classification results are more accurate for any training size. Especially, for a training set of 500 k documents, we achieve an accuracy of 93.8%. As we can see, the 500 k documents form a threshold for the classifier. For both versions of the algorithm after this point accuracy degrades. This is because there is an overfit of data, which is a classic problem in Machine Learning approaches.

4   Related Work

This work is not the first that proposes the use of social network [9,10,14,15] or mobile phone data [13] for crisis response and management. AIDR [9] is a platform that classifies tweets into user-defined, crisis-related categories in a semi-automatic way (classifiers are built using crowdsourced annotations). The work in [15] focuses on identifying timely and relevant responders for questions in social media in order to help people affected by natural calamities. The authors of [10] employ unsupervised learning techniques on textual data to extract knowledge and provide an overview of the crisis while the study in [13] correlates CDR

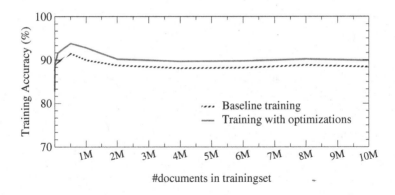

Fig. 4. Classification accuracy for different training sizes

data with rainfall levels, suggesting the potential in using cell tower activity information to improve early warning and emergency management mechanisms in case of floodings.

Contrarily, our platform focuses on combining textual as well a spatial data (GPS traces) from multiple social networks to (a) detect possible emergency POIs and to (b) provide personalized information about a crisis situation. Furthermore, our platform relies on distributed processing and storage solutions and thus proves to be scalable, while no implementation and performance information is provided for the other systems. Our platform is based on prior work of our team, which resulted in a socially enhanced recommendation system [11,12]. This has been adapted to cater the specific requirements of crisis detection and management.

5   Conclusions

In this paper we presented a storage and processing platform that is able to support applications and services that leverage the power of Big Data produced by mobile and social network users to detect and manage emergencies. Such data include spatio-temporal and textual information, which can be combined to automatically discover POIs and events that could indicate an emerging crisis of any extent, provide crisis-related information based on criteria such as location, time, sentiment or a combination of the above and infer a user's semantic trajectory during and after the emergency. Our prototype, which currently supports Facebook, Twitter and Foursquare, is able to provide query latencies of a few seconds even under heavy load, falling into the sub-second scale when executing over a 16-node cluster. Releasing an online public version of our system and testing it under real world conditions is part of our future plans.

References

1. Apache HBase. http://hbase.apache.org
2. Apache Mahout. https://mahout.apache.org/
3. Digital in 2017: Global Overview. https://wearesocial.com/special-reports/digital-in-2017-global-overview
4. Facebook Safety Check. https://www.facebook.com/about/safetycheck/
5. Facebook Stats. https://newsroom.fb.com/company-info/
6. Postgresql. http://www.postgresql.org/
7. Twitter Usage Statistics. http://www.internetlivestats.com/twitter-statistics/
8. He, Y., Tan, H., Luo, W., Mao, H., Ma, D., Feng, S., Fan, J.: MR-DBSCAN: an efficient parallel density-based clustering algorithm using mapreduce. In: ICPADS (2011)
9. Imran, M., Castillo, C., Lucas, J., Meier, P., Vieweg, S.: AIDR: artificial intelligence for disaster response. In: Proceedings of the 23rd International Conference on World Wide Web, pp. 159–162. ACM (2014)
10. Lazreg, M.B., Goodwin, M., Granmo, O.-C.: Deep learning for social media analysis in crises situations. In The 29th Annual Workshop of the Swedish Artificial Intelligence Society (SAIS) 2–3 June 2016, Malmö, Sweden, p. 31 (2016)
11. Mytilinis, I., Giannakopoulos, I., Konstantinou, I., Doka, K., Koziris, N.: MoDisSENSE: a distributed platform for social networking services over mobile devices. In: 2014 IEEE International Conference on Big Data (Big Data), pp. 49–51. IEEE (2014)
12. Mytilinis, I., Giannakopoulos, I., Konstantinou, I., Doka, K., Tsitsigkos, D., Terrovitis, M., Giampouras, L., Koziris, N.: MoDisSENSE: a distributed spatiotemporal and textual processing platform for social networking services. In: Proceedings of the 2015 ACM SIGMOD International Conference on Management of Data, pp. 895–900. ACM (2015)
13. Pastor-Escuredo, D., Morales-Guzmán, A., Torres-Fernández, Y., Bauer, J.-M., Wadhwa, A., Castro-Correa, C., Romanoff, L., Lee, J.G., Rutherford, A., Frias-Martinez, V., et al.: Flooding through the lens of mobile phone activity. In: 2014 IEEE Global Humanitarian Technology Conference (GHTC), pp. 279–286. IEEE (2014)
14. Purohit, H., Sheth, A.P.: Twitris v3: from citizen sensing to analysis, coordination and action. In: ICWSM (2013)
15. Ranganath, S., Wang, S., Hu, X., Tang, J., Liu, H.: Facilitating time critical information seeking in social media. IEEE Trans. Knowl. Data Eng. **PP**(99), 1–14 (2017)
16. Zielinski, A., Middleton, S.E., Tokarchuk, L.N., Wang, X.: Social media text mining and network analysis for decision support in natural crisis management, pp. 840–845. Proc. ISCRAM. Baden-Baden, Germany (2013)

Robotic Systems for Crisis Management

Robots in Crisis Management: A Survey

Ioannis Kostavelis$^{(\boxtimes)}$ and Antonios Gasteratos

Laboratory of Robotics and Automation Department of Production
and Management Engineering, Democritus University of Thrace,
Vas. Sophias 12, 671 32 Xanthi, Greece
{gkostave,agaster}@pme.duth.gr
http://robotics.pme.duth.gr

Abstract. Robots are flexible intelligent machines designed to replace human operators in dangerous, dirty, and dull tasks. Therefore, an apparent application of robotics is to provide assistance in managing critical situations, by carrying out tasks that might be harmful for us. Indeed, almost any police corps around the globe is nowadays equipped with a a counter improvise explosive device (IED) robot. Although counter-IED robots constitute the most famous robots for crisis management, they formulate just one of the areas that robots can be used for. An abundance of research has been performed through the last ten years in safety, security and rescue robotics. This paper attempts to review the efforts reported so far. Moreover, it attempts to systemize the robots for crisis management, in the sense that it categorizes them according to the specific use and examines which architecture would be most suitable for each case.

Keywords: Robotics · Crisis management · USAR · IED · EOD

1 Introduction

Crisis response involves great chance of risk due to unpredictable conditions occurring at the scene of the incident. The uncertainty prevailing under critical circumstances is likely to endanger the life of the rescuer workers, as well as any first responder. In order to decrease uncertainty civil protection agencies around the word have issued regulations and standard procedures to cope with catastrophic incidents. Such measures recently include the employment of advanced mechatronics and robotics systems [1]. The valuable potential of robots in crisis management has been recognized since the early beginning of the robotics era. No matter the recognition of the usefulness of systems acting autonomously in crisis scenarios the technology gap did not allow the realization but tele-operated machines [2,3], that nowadays would be characterized as primitive and obsolete. It was only until the last decade that the flourish of robotics technology, along with the research in lateral technologies –including new light and endurable materials and batteries, powerful portable computing devices, advances in computer vision–, gave thrust for an abundance of proposals

© Springer International Publishing AG 2017
I.M. Dokas et al. (Eds.): ISCRAM-med 2017, LNBIP 301, pp. 43–56, 2017.
DOI: 10.1007/978-3-319-67633-3_4

for robots that act and behave autonomously in crisis scenarios. Since 2007, it has been stated that the lack of human-robot collaboration currently presents a bottleneck to widespread use of robots in urban search and rescue (USAR) [4]. A solution to this issue is to endow contemporary robots with functionalities that would substantially enhance the efficiency and effectiveness of rescue workers and, at the same time, reduce their cognitive workload.

During the last decade two specifically organized attempts from the field of robotics gave thrust in the promotion of robotics applications for crisis management. The first one is the well known international symposium of Safety, Security and Rescue Robotics (SSRR), yearly organized since 2009, which mainly focusses on covering the design and implementation of robotic applications targeting on search and rescue, rapid screening of travelers and cargo, hazardous material handling, humanitarian demining, and detection of chemical, biological, radiological, nuclear, or explosive risks [5,6]. The second leverage factor that steered the interest of robotics community towards the field of crisis management was the DARPA Robotics Challenge (DRC). DRC is a prize competition funded by the US Defense Advanced Research Projects Agency and held from 2012 to 2015. It aimed to develop semi-autonomous ground robots that could do *"complex tasks in dangerous, degraded, human-engineered environments"* that closely resemble disaster sceneries [7]. The contributions of these two events was twofold; On the one hand, they aroused the attention of universities and research institutes on developing robotics solutions suitable for first response and crisis management and, on the other hand, it eventually led to the development of valuable cutting-edge solutions. In turn, such solutions have been adopted by the national organizations of several countries responsible for the crisis management after natural disasters and terrorism attacks. The authors of the paper at hand are well aware of the added value of these applications and aim to identify and organize herein in a systematic manner the robotics solutions developed for such application scenarios so as to ease the respective community in its future work.

2   Competitions

Dedicated infrastructures have been built to support the evaluation of the developed robots for safety, security and crisis management applications. Of such are specific arenas constructed under the scope of RoboCupRescue Robot League [8,9], the ultimate goal of which is to increase awareness of the challenges in deploying robots for emergency response applications such as urban search and rescue and bomb disposal while offering objective performance evaluations of robots operating in complex yet repeatable environments. Within RoboCupRescue Robot League particular specifications and evaluation metrics have been defined for the competitors so as to measure the abilities of the developed robots to meet the requirements of the everyday realistic hazardous events. In the Old Continent the ELROB (the European Land Robot Trial) and more recently euRathlon –part of the European Robotics League (ERL)– are a couple of other robotics competition allowing to demonstrate and compare the capabilities of robotics systems in realistic scenarios and terrains. In both competitions the participating teams aim to test the autonomy and intelligence of the involved robots in realistic mock emergency-response scenarios [10]. Moreover, Robin Murphy in 2001 marked the opening of the Disaster City® [11] a specifically designed arena that resembles to natural disaster sceneries, where first responders and robotic teams can be trained. Murphy and her team serves as a crisis response and research organization striving to direct and exploit new technology development in robotics and unmanned systems for humanitarian purposes worldwide.

3   Architectures

It is evident that the robotics design architecture depends on the application scenario and the operational environment. Consequently, several non-uniformities exist among the developed robots, exhibiting variations among the locomotion system utilized to traverse in uneven terrains or in water and variation to the perception mechanisms depending on the weather, lighting, climate and level of contamination conditions of the targeted operational environment. Yet, there are several common characteristics among the existing solutions such as remote control units suitable for teleoperation and guidance of the robot, user interphases for situation awareness and simulation tools for mission contingency mitigation strategies. Figure 1 illustrates the user interfaces of View-Finder [12] and AVERT projects [13], both comprising first response robotic solutions for hazardous and counter-terrorism situations. Note that in both control units the map of explored environment is exhibited providing thus insights to the user about the environment understanding from the robot's perspective.

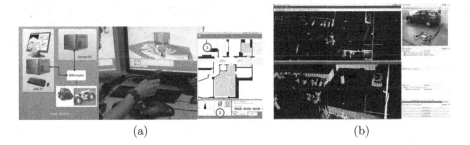

(a)   (b)

Fig. 1. The control panel of (a) View-Finder [12] and (b) the AVERT project [13]

3.1   Wheeled and Caterpillar Robots

Several mobile robots with various architectures have been developed during the last decades for safety and security applications with different locomotion systems. Specifically, wheeled mobile robots have been developed for surveillance as security guards and security officers in indoors (Cobalt Robotics [14]) and outdoors (SMP Security Robots [15]) applications utilizing omnidirectional and differential or Ackerman steering topologies, respectively. In the counter terrorism domain disarming of trapped vehicles with explosives by means of mobile robots has been extensively studied. The main objective of the developed applications include lifting trapped vehicles with explosives and carrying them away from urban and congestive environments, thus allowing their safe disposal. For this reason Tohoku University developed a series of robots, namely intelligent Cooperative Autonomous Robot Transporters (iCART) which belongs to the category of Explosive Ordnance Disposal (EOD) robots [16]. iCART II consists of four units that dock to the car, lift and transport it. The extraction trajectory is provided to the first robot, while the remaining ones estimate their respective trajectory through the interaction forces between each other. A next generation solution to the similar application was provided by the AVERT robot [13] consisted of four small sized omnidirectional *bogies* that dock to the car tires and lift the entire vehicle for its safe removal Fig. 2. AVERT acts as a plugin to existing EOD robots working together to further expedite response time, and enabling the EOD team to undertake parallel activities. Also within AVERT project a real-time path planning algorithm for dynamic obstacle avoidance in cluttered environments has been developed [17]. The obstacle detection algorithm was a specifically designed holistic solution relying on 3D vision processing and terrain learning suitable for the detection of barriers in front of the robot and has been extensively described in [18].

(a)   (b)

Fig. 2. (a) The iCART II robotic unit and (b) the next generation AVERT robotic solution, for extracting and carrying vehicles

Caterpillar robots are dedicated to maneuver in uneven terrains with existence of debris, targeting operation after a natural disaster or in antiterrorist interventions. Some representative robots for such applications are summarized herein. Specifically, the Jacobs robot [19,20] supports fire brigade during the situation assessment in a hazmat rescue drill. The robot has been tested in an operational scenario consisted of a traffic accident involving a truck transporting hazardous material and several passenger cars. In a more complex task two caterpillar robots support first responders at the European Land Robot Trial 2007 in the situation assessment after a simulated terrorist attack with NBC substances at a large public event. The two robots are supervised by only a single operator. The robots can operate fully autonomously and perform a coordinated exploration of the area [21]. Caterpillar robots are also utilized by several police departments for explosive mechanisms neutralization (see e.g. in Fig. 3(a) the one towing the deployment unit of the AVERT system [13]). Another representative paradigm of such applications is the RESCUER robot. The main objective of this robot was to improve the risk management in emergencies through mechatronic support for explosives deactivation. It comprises a multipurpose robot which aims to replaces the current highly specialized robots on deactivation or rescue, by retaining the capacity to undertake large number of missions such as intervention, sampling, and situation awareness in CBRN (chemical, biological, radiological, and nuclear) [22]. RESCUER robot is equipped with a stereoscopic system [23] protected by a transparent protective shield to be able to perceive and, thus operate in harsh environments (see Fig. 3(b)).

(a)   (b)

Fig. 3. (a) The police caterpillar robot, used for the neutralization of explosive mechanisms and (b) RESCUER robot operating in a field environment

3.2   Legged Robots

In USAR missions it is often necessary to place sensors or cameras into dangerous or inaccessible areas to obtain better situation awareness for the rescue personnel, before they enter a possibly hazardous area. This was the main objective of the development of ASGUARD robot [24]. The latter was a hybrid legged-wheeled robot able to cope with stairs and excessively rough terrain while it was navigating also fast on flat ground. Another realistic work on legged robots describing the limitations and challenges that contemporary legged robots should confront with in the aftermath of a disaster is summarized in [25]. The ATLAS robot -illustrated in Fig. 4(a)- has been developed for the DRC. During this event, the were specific tasks on which the competitive robots were tested on several challenging tasks, the most representative of which were door opening, pipes manipulation, operation uneven terrain, ladders climbing and manipulation of articulated objects. In a more recent work [26] HRP-2Kai legged robot was presented. The latter had the capability for highly autonomous operation in an unknown environments, the physical capacity of walking on uneven terrain together with a large step and a steep slope and it was adaptable to working environment and manipulation tasks. Moreover, authors in [27] developed a miniaturized five legged robot suitable to operate in cluttered environments. The robot was inspired from the sea star and its main advances was that it retained more than one learning module to control the navigation of the five legs. In a more sophisticated work [28,29] a six legged rescue robot has been developed (see Fig. 4(b)). Compared to other legged-robots, the main difference is that three-link parallel mechanisms form its legs. Accordingly, the most outstanding features of this robot is the heavy payload, the increased static stability, and the adaptation to uneven terrain. In a similar work [30], a hexapod robot has been developed equipped with wireless image sensor for real time video transmission. Its user possesses the capacity to control the robot over wireless communication and a tripod gait mechanism controls the movement of the hexapod.

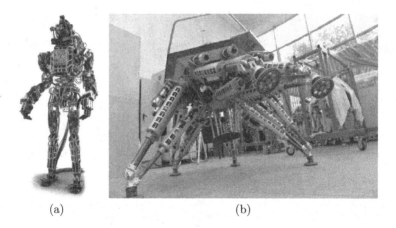

(a)   (b)

Fig. 4. (a) The ATLAS robot participated in DRC and earned the seventh (out of 16) position [25] and (b) the six legged robot capable of traversing in harsh terrain [28]

3.3 Snake Robots

Snake-like robots seem to be a very promising class of mobile robots which can fulfill most of the requirements needed for practical search-and-rescue artificial agents [31]. Serpentine robots receive much attention in the search and rescue robotics research due to their aspect ratio of cross sectional area over length and their highly redundant kinematics. These two characteristics allow such robots to operate dexterously through voids and uneven surfaces with various textures in a small space. Serpentiform movement is suitable in a variety of search and rescue missions allowing to maneuver through terrain that would have been difficult to deal with otherwise [32]. In a specific project the developed robot was coupled with a rescue canine that acted as the deployment unit of the tether serpentiform

artificial agent targeting to explore a collapsed building. The capacity of the
robot to operate in crisis situations was evaluated in the premises of TEEX
Disaster City. Moreover, the authors in [33] researched the serpentine gait of a
snake robot with a camera place on its head. They found that the camera view
stability is diminished due to the large swinging amplitude of the snake robot
head during locomotion. Aiming at improving this stability they employed an
optimization approach to modify the relative angle between the snake robot head
and its body, thus affording exploration of cluttered environments.

In [34] it is supported that most robotic search and rescue solutions involve
large, tank-like robots that use brute force to cross difficult terrain and, instead,
they proposed the Hybrid Exploration Robot for Air and Land Deployment
(HERALD), which is illustrated in Fig. 5(a). The prototype consisted of three
nimble, lightweight robots that can travel over difficult obstacles by air, but also
can travel through rubble using the snake-like component (Fig. 5(b)). For similar
rescue and survey applications the OmniTread OT-4 serpentine robot [35] was
developed. OT-4 has a unique design characteristic, that is the use of pneumatic
bellows to actuate the joints, which allow simultaneous control of position and
stiffness for each joint. The adopted architecture enables the OT-4 snake robot
to climb over obstacles that are much higher than the robot itself, propel itself
inside pipes of different diameters, and traverse even the most difficult terrain,
such as rocks or the rubble of a collapsed structure.

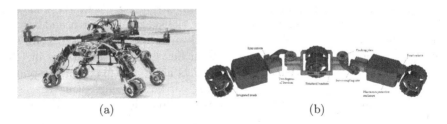

(a)   (b)

Fig. 5. The HERALD hybrid robot (a) the integrated hybrid system and (b) the HER-
ALD snake robot component. The figures have been adopted from [34]

3.4   Aerial, Marine and Other Robots

Aerial robots have attracted increased research attention during the last years,
due to major environmental disasters that emerge the immediate and remote
engagement of agents for sensing and rapid situation awareness. The already
well known Unmanned Aerial Vehicles (UAVs) earned an abundance of advancement and a representative example is the Hellenic Civil Unmanned Aerial Vehicle
(HCUAV) project which aimed to provide solutions to both illegal migration and
wildfires [36]. The proposed solution comprised a High Altitude, Long Endurance
vehicle capable of flying at 20Km altitude and was integrated with low electronics and sensors which combined with sophisticated image processing algorithms
for delivering a cost efficient platform suitable for accurate remote sensing and

surveillance. In addition, the authors in [37] designed a quadcopter robot suitable for the remote monitoring of disaster environments. The quadcopter robot is an Unmanned Aerial Vehicle or UAV that consists of four propellers and four brushless motors that function as actuator. This robot was able to fly in 30 m altitude and was equipped with a GoPro HD Hero2 camera facilitating monitoring of any places that are hard to reach. In a similar work the authors in [38] integrated a target position correction solution with an Endurance quadcopter suitable for search and rescue mission. In an effort towards advancing the performance of UAVs in disaster environments, the authors in [39] argue that a swarm is more efficient than a single unit for crisis management. Therefore, they simulated how one damaged unit would affect the entire system performance using the V-REP robotic software simulator, where swarms exhibited increased overall accuracy and reliability against one single unit. In a more collaborative work which is outlined TRADR project [40], a hybrid rescue team consisted of ground, aerial vehicles and human teams aimed to provide incremental capability improvement over the duration of a mission. A survey that summarizes the contemporary status of the advances in unmanned aerial vehicles (UAVs) for network-assisted first response to disaster management, can be found in [41].

A side effect of the great earthquake in Japan was the tsunami hazard. In the aftermath of the tsunami specific rescue team used underwater SARbot and Seamor at the port site in order to determine whether there was at least 5 meters of clearance for fishing boats to enter the port. Identifying the need to deploy robots for inspection and rescue after an earthquake, the authors in [42] developed a telescopic universal joint used for the snake-like search and rescue robot to observe the earthquake ruins and find survivors. Furthermore, counter-catastrophic robotic solutions have also been developed. The authors in [43] developed a mechatronic mechanisms able to estimate the maximum response of a building when subject to a devastating earthquake only using a sensor agent robot equipped with an accelerometer situated on an arbitrary floor of the building. Moreover, in the work presented in [44], a novel robotic system, aiming to spread a narrow gap and open a new space to release buried survivors from pressure has been developed.

4   Application Scenarios

Chemical, Biological, Radiological, Nuclear, and Explosive (CBRNE) incident response varies dramatically based upon the hazardous material, the incident size, and the response duration. Robots can facilitate response planning, maintaining situational awareness, removing responders from dangerous situations, and allowing for immediate site feedback prior to human responder entry. A summary of the application scenarios where the robots can have the role of search and rescue can be found in [45], however in the following sections we refer to the most representative applications further than SAR.

4.1 Improvise Explosive Device Disposal and Explosive Ordnance Disposal (IEDD/EOD)

During the last decades, explosive ordnance disposal teams started to use robots to access and neutralize explosive mechanisms. EOD robots became increasingly popular meaning to replace human teams in many demanding tasks, such as reconnoitering, handling and inspecting explosives, all performed directly in the dangerous field of operation [46]. It is a fundamental requirement that such robots should be designed for the coarse ambient in which they operate and have to be equipped with advanced capabilities and behaviors to enable them to operate more independently with less of a need for human intervention [47].

In June of 2007, a suspect vehicle was reported near a central London nightclub and the EOD experts used a robot to investigate the vehicle, and a large amount of fuel, gas canisters, and nails were discovered inside [48]. The Talon, a medium-tracked multipurpose robot, has been widely used in multipurpose operations including the World Trade Center attacks and the Fukushima nuclear reactor meltdown. Approximately 3, 000 Talon robots are in operation, and they are used in more than 20.000 EOD operations featuring a variety of payloads, such as night vision cameras with thermal and zoom options; chemical, biological, radioactive, and nuclear sensor packages, as well as counter-IED systems and gripper manipulators [49]. The AVERT robotic system described in [13] was designed to work as plugin to any available EOD robot, in order to deploy a swarm like robotic units to remove vehicles that block a car with explosives. AVERT robotic unit provided key operational attributes, including better situation awareness for the EOD expert teams, access paths that cannot realistically be provided by the existing EOD robots and faster and safer removal capability than is achievable with the manual methods currently available.

Authors of [50] acknowledged that landmine crisis is globally alarming since there are presently 500 million unexploded, buried mines in about 70 countries. Towards this direction they designed a robot prototype which is capable of detecting buried land mines and changing their locations, while enabling the operator to control the robot wirelessly from a distance. Under the same scope of crisis management falls also the necessity for remote radiation measurement, given the recent nuclear accidents like that in Fukushima Daiichi. Aiming to tackle that issue, the authors in [51] designed an all-terrain for remote radiation mapping inside nuclear installations. The robot was designed to operate in uneven terrain including staircases and was equipped with three onboard cameras and a retractable radiation detector.

4.2 Physical Disaster

The utilization of artificial agents in physical disaster management is an ongoing and active research topic. Rescue robots have participated in confronting at least 28 disasters in six countries, since the first deployment at the 9/11 World Trade Center collapse [11]. Their necessity to operate in situations where human access is impossible or life threatening is therefore apparent. According to [41], the response time of disaster management personnel during a natural incident is of key importance in saving the lives of those in peril. The most efficient situational awareness is achieved through aerial assessment, which is typically conducted by UAVs. This is due to the fact that UAVs can provide better insights about which structures were affected by the event, the extent of the damage, the state of the transportation infrastructure and the potential number of people influenced by the event. Therefore, the deployment of UAVs after a natural disaster is a useful tool in the hand of the first responders, providing a rapid inference of the current situation and for monitoring purposes.

Moreover, robots have already been used in earthquake situations with an exemplar paradigm the Great Eastern Japan incident [52]. Specifically, the caterpillar KOHGA3 robot has been utilized mainly for inspection in the disaster site such as a damaged building, an underground mall or in/on rubble piles where there was increased possibility for collapse. Considering the presence of fire after a natural disaster the authors in [53] developed a humanoid robot capable of pulling a fire hose while walking towards a desired position and orientation aiming to help/replace humans in dangerous situations. Considering another fire rescue robot, the work presented in [54] described a wheeled robot enhanced with hydropneumatic components capable of reaching high speeds and drive on significant roughness and terrain declivity. The authors in [55] consider that firefighting operations can be very dangerous due to low visibility, high temperature, smoke filled environments etc. and, therefore, developed an autonomous firefighting system for localized fire suppression in high and low visibility environments.

4.3   Industrial Incident or Chemical Spill

Contemporary robots also retain an active role in the prevention and intervention in the industrial incidents and chemical disasters. The authors in [56] developed a swarm robot control methodology for the chemical plume tracing (CPT) task. This concerns detection of a toxic plume and subsequent localization of the source emitter. The latter is a challenging task and requires rapid localization of the chemical emitter, thus the utilization of multiple agents is vital. Aiming to meet the actual requirements of nuclear radiation or chemical leak detection along with emergency response the authors in [57] developed a small tele-operated robot for nuclear radiation and chemical detection. The authors in [58] implemented a specific strategy for the detection of airborne chemical material on the mobile robot Arthur, which was equipped with a commercial gas sensor system. A review of mobile robots used in airborne sensing of chemical materials can be found in [59]. From a different perspective and towards the direction of protection civilians from industrial accidents, Boston Dynamics developed PETMAN robot, which is an anthropomorphic robot designed to test chemical protective clothing [60].

5   Discussion

In this paper, a short literature survey for the existing robots in crisis management has been presented. The robots have been organized with respect to their physical architecture and their operational environment. Specifically, the existing solutions have been categorized as mobile and caterpillar robotic platforms, legged, aerial and snake-like robots which have been widely used in the domain of rescue and security. Moreover, emphasis has been given in the description of the real application scenarios on which the existing robotic solutions are applying. It is concluded that various robots have been utilized in IEDD and EOD applications, natural and industrial or chemical incidents proving that artificial agents can provide significant assistance in rescue and security applications especially in scenarios where close distance operation by human is hazardous. From the above analysis it is concluded that contemporary robots are not able to manage crisis situations on their own, since on their current state they constitute devices that execute specific actions as a part of crisis management while all the robots presented in this work are used under difficult conditions as surrogate of the humans. Yet, it should be noted that fully autonomous robots for crisis management are still futuristic, since human should enter the loop to provide a complete human-robot and robot-robot collaborative solution for security and safety means. Therefore, we have only scratched the surface of the wide applications variety that artificial agents can be applied and, thus, it is anticipated that a burst of relevant solutions will make its depute in the near future.

References

1. Fink, S.: Crisis Management: Planning for the Inevitable. American Management Association, New York (1986)
2. Dixon, J.K.: The prospect of an under water naval robot. Naval Eng. J. **92**(1), 65–72 (1980)
3. Vertut, J.: Virgule- a rescue vehicle of the new teleoperator generation. In: 2nd, Conference on Industrial Robot Technology, Birmingham, England (1974)
4. Looije, R., Neerincx, M., Kruijff, G.J.M.: Affective collaborative robots for safety & crisis management in the field. In:Intelligent Human Computer Systems for Crisis Response and Management (ISCRAM 2007), Delft, Netherlands, vol. 1, May 2007
5. Community, R.: IEEE International Symposium on Safety, Security, and Rescue Robotics (2017). http://www.ieee-ras.org/conferences-workshops/fully-sponsored/ssrr
6. Guizzo, E., Ackerman, E.: The hard lessons of darpa's robotics challenge [news]. IEEE Spectr. **52**(8), 11–13 (2015)
7. Pratt, G., Manzo, J.: The darpa robotics challenge [competitions]. IEEE Robot. Autom. Mag. **20**(2), 10–12 (2013)
8. League, R.R.: Performance Metrics and Test Arenas for Autonomous Mobile Robots (2017). https://www.nist.gov/el/intelligent-systems-division-73500/performance-metrics-and-test-arenas-autonomous-mobile-robots
9. Balakirsky, S., Carpin, S., Visser, A.: Evaluating the robocup 2009 virtual robot rescue competition. In: Proceedings of the 9th Workshop on Performance Metrics for Intelligent Systems, pp. 109–114. ACM (2009)

10. Schneider, F.E., Wildermuth, D.: Comparison of the European disaster response robotics competitions Eurathlon and Elrob. In: UDC 681.518 (04) INTERACTIVE SYSTEMS: Problems of Human-Computer Interaction.-Collection of scientific papers.- Ulyanovsk: USTU, 2015.- 306 p., 23 (2015)

11. Murphy, R.R., Tadokoro, S., Kleiner, A.: Disaster robotics. In: Siciliano, B., Khatib, O. (eds.) Springer Handbook of Robotics, pp. 1577–1604. Springer, Cham (2016). doi:10.1007/978-3-319-32552-1_60

12. Baudoin, Y., Doroftei, D., De Cubber, G., Berrabah, S.A., Pinzon, C., Warlet, F., Gancet, J., Motard, E., Ilzkovitz, M., Nalpantidis, L., et al.: View-finder: robotics assistance to fire-fighting services and crisis management. In: International Workshop on Safety, Security & Rescue Robotics, pp. 1–6. IEEE (2009)

13. Amanatiadis, A., Charalampous, K., Kostavelis, I., Birkicht, B., Andel, B., Meiser, V., Henschel, C., Baugh, S., Paul, M., May, R., et al.: Autonomous vehicle emergency recovery tool: a cooperative robotic system for car extraction. J. Field Robot. **33**, 1058–1086 (2015)

14. Robotics, C.: A mobile robot for security applications in indoors environments (2017). https://www.cobaltrobotics.com/

15. Robots, S.: SMP Security Roobots for outdoors applications. https://smprobotics.com/security_robot/

16. Kashiwazaki, K., Yonezawa, N., Endo, M., Kosuge, K., Sugahara, Y., Hirata, Y., Kanbayashi, T., Suzuki, K., Murakami, K., Nakamura, K.: A car transportation system using multiple mobile robots: ICART II. In: International Conference on Intelligent Robots and Systems, pp. 4593–4600. IEEE (2011)

17. Charalampous, K., Kostavelis, I., Amanatiadis, A., Gasteratos, A.: Real-time robot path planning for dynamic obstacle avoidance. J. Cell. Autom. **9**, 195–208 (2014)

18. Kostavelis, I., Gasteratos, A., Boukas, E., Nalpantidis, L.: Learning the terrain and planning a collision-free trajectory for indoor post-disaster environments. In: International Symposium on Safety, Security, and Rescue Robotics, pp. 1–6. IEEE(2012)

19. Birk, A., Carpin, S.: Merging occupancy grid maps from multiple robots. Proc. IEEE **94**(7), 1384–1397 (2006)

20. Rooker, M.N., Birk, A.: Multi-robot exploration under the constraints of wireless networking. Control Eng. Pract. **15**(4), 435–445 (2007)

21. Birk, A., Schwertfeger, S., Pathak, K.: A networking framework for teleoperation in safety, security, and rescue robotics. Wireless Commun. **16**(1), 6–13 (2009)

22. Guzman, R., Navarro, R., Ferre, J., Moreno, M.: Rescuer: development of a modular chemical, biological, radiological, and nuclear robot for intervention, sampling, and situation awareness. J. Field Robot. **33**, 931–945 (2015)

23. Gasteratos, A.: Tele-autonomous active stereo-vision head. Int. J. Optomechatronics **2**(2), 144–161 (2008)

24. Eich, M., Grimminger, F., Kirchner, F.: A versatile stair-climbing robot for search and rescue applications. In: International Workshop on Safety, Security and Rescue Robotics, pp. 35–40. IEEE (2008)

25. DeDonato, M., Dimitrov, V., Du, R., Giovacchini, R., Knoedler, K., Long, X., Polido, F., Gennert, M.A., Padır, T., Feng, S., et al.: Human-in-the-loop control of a humanoid robot for disaster response: a report from the darpa robotics challenge trials. J. Field Robot. **32**(2), 275–292 (2015)

26. Kaneko, K., Morisawa, M., Kajita, S., Nakaoka, S., Sakaguchi, T., Cisneros, R., Kanehiro, F.: Humanoid robot hrp-2kai-improvement of hrp-2 towards disaster response tasks. In: International Conference on Humanoid Robots, pp. 132–139. IEEE (2015)

27. Anam, K., Effendi, R., Santoso, A., Jazidie, A., et al.: Autonomous five legs rescue robot navigation in cluttered environment. Acad. Res. Int. Savap **2**(1) (January 2012). ISSN-L: 2223-9553, ISSN: 2223-9944
28. Chai, X., Gao, F., Pan, Y., Xu, Y.l.: Autonomous gait planning for a hexapod robot in unstructured environments based on 3D terrain perception. In: The 14th International Federation for the Promotion of Mechanism and Machine Science World Congress (IFToMM), Taipei, Taiwan (2015)
29. Pan, Y., Gao, F., Qi, C., Chai, X.: Human-tracking strategies for a six-legged rescue robot based on distance and view. Chin. J. Mech. Eng. **29**(2), 219–230 (2016)
30. Itole, D.A., Patil, B., Bhat, P., Sawant, S.: Hexapod for surveilling in disaster situations and un-even terrain navigation. Autom. Auton. Syst. **8**(4), 93–96 (2016)
31. Hirose, S., Fukushima, E.F.: Snakes and strings: new robotic components for rescue operations. Int. J. Robot. Res. **23**(4–5), 341–349 (2004)
32. Wright, C., Johnson, A., Peck, A., McCord, Z., Naaktgeboren, A., Gianfortoni, P., Gonzalez-Rivero, M., Hatton, R., Choset, H.: Design of a modular snake robot. In: International Conference on Intelligent Robots and Systems, pp. 2609–2614. IEEE (2007)
33. Au, C., Jin, P.: Investigation of serpentine gait of a snake robot with a wireless camera. In: International Conference on Mechatronic and Embedded Systems and Applications, pp. 1–6. IEEE (2016)
34. Latscha, S., Kofron, M., Stroffolino, A., Davis, L., Merritt, G., Piccoli, M., Yim, M.: Design of a hybrid exploration robot for air and land deployment (herald) for urban search and rescue applications. In: International Conference on Intelligent Robots and Systems, pp. 1868–1873. IEEE (2014)
35. Borenstein, J., Borrell, A.: The omnitread ot-4 serpentine robot. In: International Conference on Robotics and Automation, pp. 1766–1767. IEEE (2008)
36. Amanatiadis, A., Bampis, L., Karakasis, E.G., Gasteratos, A., Sirakoulis, G.: Real-time surveillance detection system for medium-altitude long-endurance unmanned aerial vehicles. Concurrency Computat.: Pract. Exper., 1–14 (2016, in press). Published online in Wiley InterScience (www.interscience.wiley.com). doi:10.1002/cpe
37. Sudarma, M., Swamardika, I.A., Pratama, A.M.: Design of quadcopter robot as a disaster environment remote monitor. Int. J. Electr. Comput. Eng. **6**(1), 188 (2016)
38. Zou, J.T., Pan, Z.Y., Zhang, D.L., Zheng, R.F.: Integration of the target position correction software with the high endurance quadcopter for search and rescue mission. In: Applied Mechanics and Materials, vol. 764, pp. 713–717. Trans Tech Publ (2015)
39. Yie, Y., Solihin, M.I., Kit, A.C.: Development of swarm robots for disaster mitigation using robotic simulator software. In: Ibrahim, H., Iqbal, S., Teoh, S.S., Mustaffa, M.T. (eds.) 9th International Conference on Robotic, Vision, Signal Processing and Power Applications. LNEE, vol. 398, pp. 377–383. Springer, Singapore (2017). doi:10.1007/978-981-10-1721-6_41
40. Kruijff-Korbayová, I., Colas, F., Gianni, M., Pirri, F., Greeff, J., Hindriks, K., Neerincx, M., Ögren, P., Svoboda, T., Worst, R.: Tradr project: long-term human-robot teaming for robot assisted disaster response. KI-Künstliche Intelligenz **29**(2), 193–201 (2015)
41. Erdelj, M., Natalizio, E., Chowdhury, K.R., Akyildiz, I.F.: Help from the sky: leveraging UAVS for disaster management. IEEE Pervasive Comput. **16**(1), 24–32 (2017)
42. Zhao, L., Sun, G., Li, W., Zhang, H.: The design of telescopic universal joint for earthquake rescue robot. In: Asia-Pacific Conference on Intelligent Robot Systems, pp. 62–66. IEEE (2016)

43. Mita, A., Shinagawa, Y.: Response estimation of a building subject to a large earthquake using acceleration data of a single floor recorded by a sensor agent robot. In: SPIE Smart Structures and Materials+ Nondestructive Evaluation and Health Monitoring, International Society for Optics and Photonics, p. 90611G (2014)

44. Guowei, Z., Bin, L., Zhiqiang, L., Cong, W., Handuo, Z., Weijian, H., Tao, Z., et al.: Development of robotic spreader for earthquake rescue. In: IEEE International Symposium on Safety, Security, and Rescue Robotics, pp. 1–5. IEEE (2014)

45. Humphrey, C.M., Adams, J.A.: Robotic tasks for chemical, biological, radiological, nuclear and explosive incident response. Adv. Robot. **23**(9), 1217–1232 (2009)

46. Murphy, R.R., Kleiner, A.: A community-driven roadmap for the adoption of safety security and rescue robots. In: International Symposium on Safety, Security, and Rescue Robotics, pp. 1–5. IEEE (2013)

47. Burke, J.L., Murphy, R.R., Coovert, M.D., Riddle, D.L.: Moonlight in miami: field study of human-robot interaction in the context of an urban search and rescue disaster response training exercise. Hum. Comput. Interact. **19**(1–2), 85–116 (2004)

48. Kohlmann, E.F.: "homegrown" terrorists: theory and cases in the war on terror's newest front. Ann. Am. Acad. Polit. Soc. Sci. **618**(1), 95–109 (2008)

49. Danna, V.: A new generation of military robots. Res. Appl. AI Comput. Animat. **7**, 2–5 (2004)

50. Bharath, J.: Automatic land mine detection and sweeper robot using microcontroller. Int. J. Mech. Eng. Robot. Res. **4**(1), 485 (2015)

51. Saini, S., Sarkar, U., Ray, D., Badodkar, D., Singh, M.: All terrain robot for remote radiation measurement. In: 31st International Conference on Advances in Radiation Measurement Systems and Techniques: Abstract Book (2014)

52. Matsuno, F., Sato, N., Kon, K., Igarashi, H., Kimura, T., Murphy, R.: Utilization of robot systems in disaster sites of the great Eastern Japan earthquake. In: Yoshida, K., Tadokoro, S. (eds.) Field and Service Robotics. STAR, vol. 92, pp. 1–17. Springer, Heidelberg (2014). doi:10.1007/978-3-642-40686-7_1

53. Ramirez-Alpizar, I.G., Naveau, M., Benazeth, C., Stasse, O., Laumond, J.P., Harada, K., Yoshida, E.: Motion generation for pulling a fire hose by a humanoid robot. In: 16th International Conference on Humanoid Robots (Humanoids), pp. 1016–1021. IEEE (2016)

54. Bartnicki, A., Łopatka, M.J., Muszyński, T., Rubiec, A.: Stiffness evaluation of fire rescue robot suspension with hydropneumatic components. In: Solid State Phenomena, vol. 210, pp. 301–308. Trans. Tech. Publ. (2014)

55. McNeil, J.G., Lattimer, B.Y.: Autonomous fire suppression system for use in high and low visibility environments by visual servoing. Fire Technol. **52**(5), 1343–1368 (2016)

56. Zarzhitsky, D., Spears, D.F., Spears, W.M.: Distributed robotics approach to chemical plume tracing. In: International Conference on Intelligent Robots and Systems, pp. 4034–4039. IEEE (2005)

57. Qian, K., Song, A., Bao, J., Zhang, H.: Small teleoperated robot for nuclear radiation and chemical leak detection. Int. J. Adv. Rob. Syst. **9**(3), 70 (2012)

58. Lilienthal, A., Zell, A., Wandel, M., Weimar, U.: Sensing Odour sources in indoor environments without a constant airflow by a mobile robot. In: International Conference on Robotics and Automation. vol. 4, pp. 4005–4010. IEEE (2001)

59. Lilienthal, A.J., Loutfi, A., Duckett, T.: Airborne chemical sensing with mobile robots. Sensors **6**(11), 1616–1678 (2006)

60. Nelson, G., Saunders, A., Neville, N., Swilling, B., Bondaryk, J., Billings, D., Lee, C., Playter, R., Raibert, M.: Petman: a humanoid robot for testing chemical protective clothing. J. Robot. Soc. Jpn. **30**(4), 372–377 (2012)

ROLFER: An Innovative Proactive Platform to Reserve Swimmer's Safety

Eleftherios Lygouras[1(✉)], Antonios Gasteratos[1],
and Konstantinos Tarchanidis[2]

[1] Department of Production and Management Engineering,
Democritus University of Thrace, Xanthi, Greece
elygoura@ee.duth.gr, agaster@pme.duth.gr
[2] Eastern Macedonia and Thrace Institute of Technology, Kavala, Greece
ktarch@teikav.edu.gr

Abstract. The major problem of an emergency situation is the immediate reaction and the rescue assistance when people call for help. The response time is a vital factor in search and rescue (SAR) operations, owed to the fact that potential delays may have dramatic and even lethal results. Moreover, the safety of rescue workers is another major issue and must be ensured in any circumstance. SAR operations can greatly benefit from the use of Unmanned Aerial Vehicles (UAV) to reduce the time to detect the victim and provide him/her invaluable assistance. UAVs are nimble, quick-moving and can be easily programmed to exhibit autonomous behaviors. Thus, they are able to operate in difficult environments and under circumstances onerous for humans to cope with. This paper presents ROLFER (RObotic Lifeguard For Emergency Rescue), a completely autonomous robotic aerial system for immediate provision of rescue and life-saving services, which does not require human-in-the-loop for its navigation. It aims to provide fast reaction in case of an emergency under adverse environmental conditions. ROLFER is currently studied within the framework of the specific use case to provide one or more auto-inflatable floats to a distressed swimmer, thus, significantly diminishing the response time until the arrival of rescue workers. ROLFER's architecture and its elements are introduced in detail here; these include the drone and the airborne equipment, the communication link with the base station, the controlling computer and its related software as well as the wearable equipment carried by the people to be supervised and the related software developed. The purpose of the proposed system is to enhance SAR teams operational capabilities in emergency situations in aquatic environment with adverse weather and sea conditions.

Keywords: Robotic lifeguard · Emergency rescue · Autonomous UAV · Aerial rescue system

1 Introduction

In any crisis event the need for immediate support to the victims is vital. This is due to the fact that the prospects of survival are reduced as time advances. The success of SAR operations relies on how fast they are planned and executed [1, 10]. A successful SAR

© Springer International Publishing AG 2017
I.M. Dokas et al. (Eds.): ISCRAM-med 2017, LNBIP 301, pp. 57–69, 2017.
DOI: 10.1007/978-3-319-67633-3_5

operation should find, assist and rescue human beings in distress as quick as possible, in order to mobilize any contribution victims might still be capable of making to save their own lives while they are still in position to do so [1, 2]. However, even heavily trained rescue teams may take a long time to get to an incident, although they often even take human risks during such operations in order to be able to locate and recover victims quickly. Especially, when it comes to a swimmer the surveillance of large areas is almost impossible. For this reason, there is an emerging need of a fully automated rescue system capable of accurately detecting the distressed swimmer's position and automatically act, in order to direct a completely autonomous vehicle towards him to deliver lifeguard floats, until the arrival of the rescue team on the spot. Although such an aid vehicle can be either an autonomous surface vehicle (ASV) or a UAV one, the advancement of aerial systems during the last decade and the plethora of available solutions make the second option more favorable [1, 2, 10]. Moreover, aerial rescue systems can serve as a precious life-saving technology for urgent situations. They can significantly reduce operation times, thus potentially save lives in offshore and near-shore SAR operations, as they can operate in dangerous scenarios and under adverse environmental conditions and may increase rescue teams safety [1, 10]. Researchers around the world have come up with practical technologies incorporating UAVs to save lives, especially near-shore and in man overboard situations [2, 4]. With autonomous UAVs, one or more life buoys can be provided on scene during the critical moments for the distressed swimmers to cling to and wait for rescue. Thus, it may provide an alternative to lifeguard crew and give more humans a chance of survival [1, 3].

1.1   Related Research

Given the impressive advancement in robotics technology, one would expect that unmanned search and rescue (USAR) operations would have been common. Yet, no matter that researchers, companies and governmental organizations around the globe attempt to change that fact there is still a long way to go [1–3]. Nowadays, there are a lot of market available robotic systems for SAR operations, which contribute significantly to this direction. The major disadvantage of the existing aerial rescue systems is that none of them is completely autonomous and require human-in-the-loop for their navigation [3–7, 9].

Notable research has been conducted towards the direction of saving a person who has gone overboard [1, 10]. A noteworthy approach is the complex USAR system AGaPaS (Autonomous Galileo-supported Person rescue at Sea) [6], developed by research consortium. Given such an incident, AGaPaS aims to perform rapidly, with the aim to reduce the rescue time. Apparently, AGaPaS aims for crews working on special vessels or offshore platforms, therefore, they should be equipped with a specific rescue vest, which allow to locate the crew member in case an overboard accident happens [6]. By introducing UAVs to tackle this problem, the AGaPaS system could be further improved. An emergency flotation device providing system has been proposed by RTS Ideas, suitable for distressed swimmers. The system included a UAV, namely the PARS [7, 8]. AUXdron Lifeguard is another drone, exclusively design and implemented for emergency situations in water environments. It diminished the response time, by providing the swimmer with two auto-inflatable life preservers until rescue worker's arrival [9]. Yet, AUXdron is neither completely autonomous and requires human operator for its navigation. Another notable research to this direction, is the Ryptide project. Ryptide project also focuses on carrying life-rings to rescue distressed swimmers. More specifically, Ryptide, is actually an equipment designated to be installed onto UAV to carry an inflatable life-ring. After the drone reaches above the swimmer, the deflated life-ring is released and self-inflates as it hits the water [11]. Two more impressive inventions as far as concern aerial rescue systems, are (a) an octocopter created by two PhD students to assist distressed swimmers by carrying a floating device using magnets and is released by remote control [12], which won the top prize on a student prototype competition, and (b) an octocopter invented by students of Ajman University in association with Microsoft and has night vision capabilities. The main disadvantage of this aerial rescue system is that it can only be operated by lifeguards or rescue team members via radio control.

Hence, there is an arising necessity for a fully automated aerial rescue system able to accurately locate a distressed swimmer and act automatically, so as to direct a completely autonomous UAV towards him to deliver one or more lifeguard floats, until the appearance of the rescue team [1, 2, 10]. In this paper, we present a completely autonomous robotic aerial system, namely the Robotic Lifeguard For Emergency Rescue (ROLFER). The proposed system aspires to provide immediate rescue and life-saving services to swimmers in peril. It has been designed to perform and navigate fully autonomously, guided only by the position emitted by the victim. ROLFER's architecture and its building blocks are presented and discussed in detail here. The proposed system is targeting to enhance SAR teams operational capability in emergency situations in aquatic environment.

2 ROLFER's Description

Taking advantage of today's robotic technology, the main component of ROLFER system is a completely autonomously moving UAV (drone), which can provide access to beaches or to non-accessible shores, as well as anywhere on the land. In this way, rescue teams may have a great advantage in terms of immediate reaction and rescue assistance in an emergency situation. ROLFER's UAV is able to navigate completely autonomously above the water level, carrying a life raft, which is automatically inflated close to the victim.

ROLFER's architecture is graphically illustrated in Fig. 1. It includes the drone and the airborne equipment, the communication link with the base station, the controlling computer and its related software, a mobile phone or a tablet and the related Android applications for the base station, as well as the wearable equipment carried by the people to be supervised and also the related Android application developed for the swimmer.

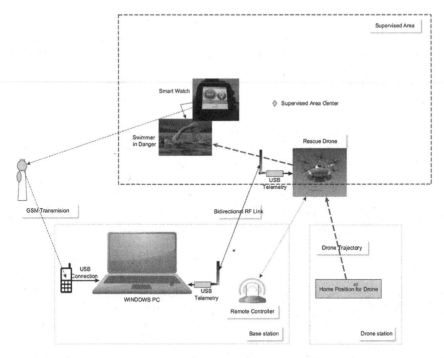

Fig. 1. The proposed aerial rescue system architecture.

More specifically, the proposed architecture also includes the following modules:

1. The *drone station*, where the operating drone carrying the lifeguard or any other suitable equipment can be landed and take-off and may contain also its battery wireless charger.
2. The *base station*, including the main controlling computer (a Windows PC), an Android mobile phone or a tablet interconnected through USB connection, a USB telemetry link to communicate with the drone and the remote controller for the manual operation of the drone, in case it is apparent.
3. The swimmer is supposed to be inside a *supervised area*. This area is specified by its *Supervised Area Center* and its length and width are specified by the deviations above this center in Latitude and Longitude. Thus, its length and width are specified as (RefLat + DevLat, RefLat − DevLat) and (RefLong + DevLong, RefLong − DevLong) respectively, where RefLat and RefLong are the coordinates of the *Supervised Area Center*. The deviations for Latitude and Longitude are specified also by the system administrator. The wearable *smart watch* for the supervised person (a swimmer in our case study), runs a special developed application on Android operating system and has almost the same capabilities as a smart phone.

Given the fact that the proposed system should run fully autonomous in extremely delicate and critical situations, the following limitations have been considered along its development:

(a) The detection/tracking of the location of the person-target with a centimeter accuracy, using GPS-based geographic coordinate tracking system, GPS-Glonass as well as the DGPS differential geo-location system.

(b) The emission and the transmission of the distress signal at regular intervals to the base station, via smart-watch with the following specifications:

- Accurate Global Positioning System
- Global System for Mobile communication capability
- Bluetooth capability
- Adequate internal memory & loading of external applications
  The purpose of multiple distress signal transmissions is to keep the base station informed of possible changes of person's position continuously over time, e.g. in the case of a swimmer or surfer drifting from races.

(c) Receiving and controlling of the distress signal and the accurate person's position data from a Tablet laptop or Laptop base station with a built-in GPS system to activate the autonomous unmanned aerial vehicle.

(d) Data transmission via specific Android application to the autonomous UAV, as well as the apparent commands for auto take-off and return-to-home operations.

(e) The dispatch of all the proper rescue equipment to the person via a completely autonomous UAV with the following specifications:

- Accurate Global Position System
- Auto Take-off & Auto Return Home
- Payload carrying & release
- Camera carrying & teleoperation
- Open source software for completely autonomous flight
- Low cost
- Use and exploitation of the Geographical Coordinate Differential System - DGPS-method.

2.1   Functionality

ROLFER is based on the concept of receiving the swimmer's position via a distress signal transmitted by him/her and respond to him/her by sending automatically a completely autonomous UAV (drone), carrying all the proper equipment. Owing to the fact that human lives might depend on the proposed system, ROLFER should avail the swimmer of the required functionality, yet with minimum chance of failure. Thus, its complexity should be kept as low as possible, leading to the serial process depicted in Fig. 2. Upon activation by the swimmer the Android Intelligent Phone Watch (AIPW) submits, through the mobile service provider, a distress signal to the base station. The activation procedure is very fair for the swimmer and it only involves intensive (for more than 3 s) pushing the "HELP" button on AIPW (see Fig. 3). Once the distressed signal has been produced, a text message is sent to the base station of the platform, informing the lifeguard or the rescue workers that a swimmer's life might be at risk. A computer or a tablet in the base station receives this text message that includes the geographical coordinates of the incident. After validating and decomposing the

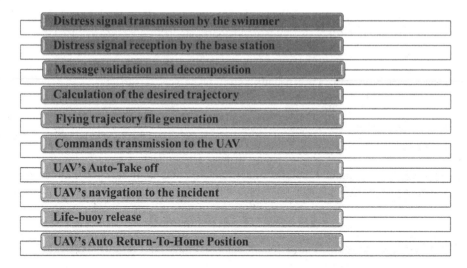

Fig. 2. Rescue process description.

received text message, an internal base station's application, plans the path of the autonomous UAV, while generating the aerial's vehicle mission file. The exact trajectory is planned based on the drone's home position, the intermediate points specified by the system administrator, considering the possible obstacles, for example people on the beach or other restrictions, and the received coordinates of the swimmer. The intermediate points can be provided as a series of successive positions. The number of intermediate points can be from zero to any desired number. The target position of the swimmer in danger is the final point to the autopilot of the drone. Next the commands are transmitted to the aerial vehicle. The base station's software runs at the same time on its computer and via a telemetry-RF link transits all the necessary commands to the aerial vehicle. These include the actions to: arm the motors; check for GPS satellites; auto take off; navigate towards the incident; detect/track the exact swimmer's position; release one or more auto-inflatable floats to him/her; auto return-to-land.

Fig. 3. Screenshot of user's application for the distressed signal transmission.

Therefore, when the UAV finds itself above the incident location, it will release the float to the distressed swimmer. Then, the swimmer can grasp the life preserver, which allows critical response until the arrival of the rescue workers. At the same time the base station sends an SMS to a ground authority, viz. the Coast Guard, with the incident GPS position, using a mobile phone for the communication between the land base station and the authorities. Once the drone has completed its mission, it auto-returns and auto-lands in the base station (Fig. 4).

Fig. 4. ROLFER's base station and UAV.

2.2   Rescue System Parameters Adjustments

(a) The system administrator
The administrator is the person supervising the rescue system and can insert and modify any operational parameter on the wearable equipment for the user (a swimmer in our case study), on the mobile phone or tablet of the base station or even the software running on the controlling computer of the base station. When entering into each of the above three applications, the administrator must be authorized. The smart watch must be suitably set-up beforehand. In the user's application, the administrator must initially insert the password and then the working parameters. The first group of parameters include the reference coordinates of the supervised area (RefLat, RefLong, RefAlt) which determine the center of the supervised area and the deviation for Lat and Long, DevLat and DevLong. In this way, (Fig. 5) the supervised area is set as a rectangular determined by the coordinates (RefLat ± DevLat and RefLong ± DevLong). The center of the area can be inserted either manually, if the reference coordinates are already known, by going in the reference point and saving the coordinates acquired by the GPS module of the smart watch. A second group of parameters include the number of messages to be transmitted in the case of emergency, the time duration between message transmission, the phone numbers of the messages destinations, a PIN number for identification of the user by the receiving base station etc. Finally, the administrator defines also the password for the user, which will use the application hereafter.

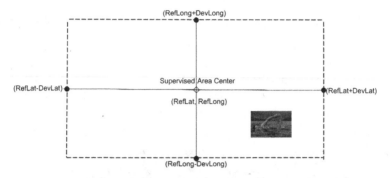

Fig. 5. Definition of the supervised area.

(b) The user's application

Although smart watches are yet a new consumer electronics product, they show a tendency to hit the mainstream. The Android Wear is a common operating system for such devices. Smart watches normally maintain an almost constant wireless connection to a mobile device –usually a smart phone–, although stand alone versions exist that do not need to pair with a smart phone. They can run applications created by several developers right on the watch, including message composing and sending, health and fitness parameters observations, navigation, to name a few. Since most smart watches are based on open software platforms, the developer community can add new and innovative apps that can increase their functionality. Their lightweight and easiness of use are some of the reasons that such an apparatus is suitable for the proposed application. Moreover, all smart watches claim some degree of waterproof: some models resist underwater up to 1 m for 30 min, while some others claim up to 100 m. After completing lab tests on some of the newest smart watches satisfying the minimum technical requirements we concluded to the more suitable for our application.

The developed application runs in Android environment and requires a minimum of 10 MB of memory. While the user is inactive, an algorithm running in the developed application improves the GPS coordinates acquired accuracy, by implementing a simple moving average (SMA) algorithm. The differential GPS algorithm has been also implemented. In a simple moving average algorithm (or equally weighted running mean) for a n-value sample $p_M, p_{M-1}, ..., p_{M-(n-1)}$ (where p_M is the last value received and n is the depth of averaging), the formula for the $\bar{p}_{SM}$ is:

$$\bar{p}_{SM} = \frac{p_M + p_{M-1} + \ldots + p_{M-(n-1)}}{n} = \frac{1}{n}\sum_{i=0}^{n-1} p_{M-i} \tag{1}$$

While calculating the average of successive values, when a new value comes into the sum an old value drops out, and thus a full summation each time is unnecessary for this simple case:

$$\bar{p}_{SM} = \bar{p}_{SM,prev} + \frac{p_M}{n} - \frac{p_{M-n}}{n} \tag{2}$$

The period selected depends on the type of movement of interest, such as short, intermediate, or long-term. In our case assuming a slowly moving swimmer, the depth of averaging has been selected as $n = 5$, thus providing satisfying accuracy of about ± 0.5 m in surface coordinates average values. The user GPS apparatus has been adjusted in calculating and refreshing the actual coordinates every 1 s.

The user wearing the smart watch, after inserting the password provided by the administrator, is capable to initialize a rescue operation upon pressing the "HELP" button presented on his/her smart watch. This fires a series of operations. First a text message is composed including the user's PIN, its coordinates and other apparent elements. Then a counter is triggered also counting the number of the messages send. The process is completed when the specified number of transmissions has been reached. The user has also the capability to cancel the rescue operation by pressing the next CANCEL button. In this case, the drone controller will then issue a Return To Launch (RTL) command to it, thus cancelling the rest of the operation.

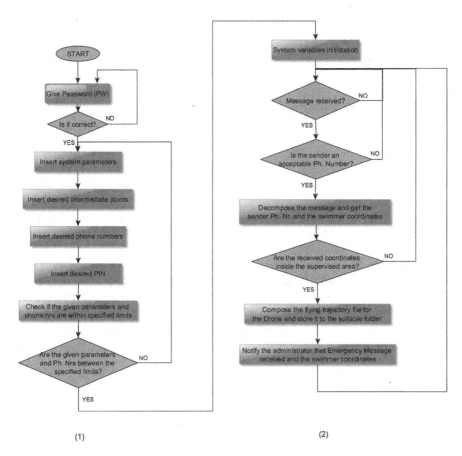

(1)   (2)

Fig. 6. Block diagram for the base station application.

(c) The base station application

A second Android application has been developed for the mobile phone or tablet on the system base station. The administrator can insert and modify also the parameters for the base station application. This application is running on an Android mobile phone or a tablet interconnected through USB connection to the main controlling computer. According to the Fig. 6, after a correct password has been provided, the administrator is capable of inserting a number of working parameters, as for example the supervised area reference coordinates, the coordinates of possible intermediate points for the drone to follow before it reaches the final swimmer coordinates, the acceptable phone numbers, the accepted PIN for the user and so on. In every parameter insertion process, either in the user's application, or in the base station one, the given parameters are preliminary checked to match specific criteria.

Next the system is set to the waiting state. In this situation the application is looking for a message received by anyone of the acceptable phone numbers. If the PIN extracted from the text field is acceptable also and if the coordinates contained in the same field are originated from the supervised area, the application produces the flying trajectory file and stores it in a special folder. In this folder the controlling computer looks at discrete time intervals and if such file is detected it issues the take-off command to the drone.

3 Simulations Results

The emergency rescue system described above has been already implemented. In order to avoid dangerous direct testing on the real system, that could lead to undesirable crashes, preliminary simulations tests are apparent. Thus the base station software of Mission Planner has been adopted. It is an open source software allowing the simulation and flight of multicopters, planes and rovers. The drone flight is programmed via scripting. The simulation tests were carried out at the Department of Electrical and Computer Engineering of Democritus University of Thrace. They aim to test, define and demonstrate the completely autonomous deployment of the aerial rescue system namely ROLFER. In Fig. 7(a), the simulated flight of the vehicle is shown during the first period where it passes through the intermitted point 1 at $Alt = 10$ m, reaches the distress signal point, from where the emergency signal received and releases the appropriate equipment at $Alt = 2.2$ m. Then it is taking-off again at the predefined altitude $Alt = 10$ m and returns through the intermitted point 2 to the launch point (base station), where it is landed (b).

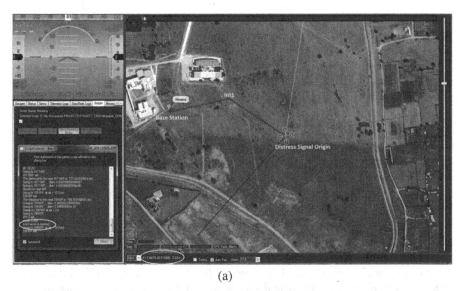

(a)

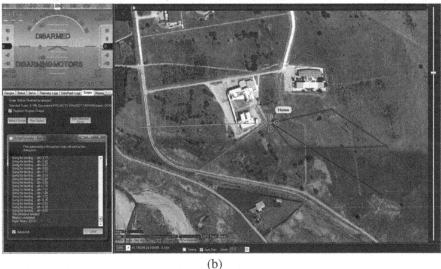

(b)

Fig. 7. Simulated flight of autonomous UAV during the first period where it passing through the intermitted point 1 (a) and UAV's return through the intermitted point 2 to the base station (b).

4 Conclusions

UAVs can avail SAR operations with critical support, yet any parameter that potentially may affect the flight of a such a rescue system should be properly taken into account. Such parameters include the quality of sensory operation, energy limitations, environmental factors or digital protection restrictions between the aerial vehicle and the rescue-worker teams. In this paper, we have presented a completely autonomous aerial rescue integrated system for life-saving services, namely the ROLFER system. The Key Performance Indicators in our project are to: (i) be fully autonomous, i.e. without requiring to include human-in-the-loop and (ii) operate as rapid as possible, i.e. to minimize the time interval the UAVs needs to provide invaluable assistance on a human in case of emergency. Our preliminary results show the importance of exploiting as much as possible the critical response time as well as other important parameters during a SAR operation. In the first preliminary tests the measured GPS accuracy was in the order of ±0.5 m. Although it is quite satisfactory, this figure must be improved. In our future work we intend to study and implement a suitable computer vision algorithm to improve the UAV's accuracy, so to release the life raft as close as possible to the victim. Specifically, the system should not release the float, until the victim is located by means of the onboard camera. The intended computer vision algorithm would track the swimmer and produces commands to the vehicle's controller, so as the UAV to move to the appropriate position above the swimmer. When the error on the image plane will be zeroed, the float will be released. Overall, it seems that aerial vehicles technology looks very promising for SAR operations and especially for the particular case of distressed swimmers. Therefore, our future work also includes the assessment of the system in real conditions, in order to prove its usability and usefulness.

Acknowledgments. The authors of the paper would like to thank Stavros Niarchos Foundation as well as the Eastern Macedonia and Thrace Institute of Technology, as ROFER project has been founded as a fellowship with their exclusive support.

References

1. Yeong, S.P., King, L.M., Dol, S.S.: A review on marine search and rescue operations using unmanned aerial vehicles. Int. J. Mech. Aerosp. Ind. Mech. Manuf. Eng. 9(2), 396–399 (2015). World Academy of Science, Engineering and Technology
2. Bennett, M.: Rescue in all realms, safety missions in europe expand past the air, unmanned systems (2015)
3. De Cubber, G., et al.: The EU-ICARUS project: developing assistive robotic tools for search and rescue operations. In: IEEE, Safety, Security, and Rescue Robotics (SSRR) (2013)
4. Matos, A., Silva, E., Cruz, N., Alves, J.C., Almeida, D., Pinto, M., Martins, A., Almeida, J., Machado, D.: Development of an unmanned capsule for large-scale maritime search and rescue. INESC TEC - Institute for Systems and Computer Engineering of Porto (2013)
5. Perez-Alberti, A., Trenhaile, A.: An initial evaluation of drone-based monitoring of boulder beaches in Galicia, north-western Spain. Earth Surf. Process. Landforms **40**, 105–111 (2014)
6. Kurowski, M., Lampe, B.P.: Agapas: a new approach for search-and-rescue-operations at sea. Proc. Inst. Mech. Eng. Part M J. Eng. Maritime Environ. **2**, 156–165 (2014)
7. RTS Ideas, Pars Drone (2014)
8. Kelion, L.: Iran Develops Sea Rescue Drone Prototype in Tehran, BBC, November 13 (2013) http://www.bbc.com/news/technology-24929924
9. General Drones: AUXdron Lifeguard. http://www.generaldrones.es/axdron_lifeguard_drone/

10. Waharte, S., Trigoni, N.: Supporting search and rescue operations with UAVs. In: IEEE International Conference on Emerging Security Technologies (EST), 6–7 September (2010)
11. Stu, R.: Project Ryptide: Drone Flies Life-Rings to Distressed Swimmers. http://www. gizmag.com/project-ryptide/35437/. Accessed 6 Jan 2015
12. University of Wollongong, Life-Saving Drone Takes out Innovation Competition (2014)

Identifying Hazardous Emerging Behaviors in Search and Rescue Missions with Drones: A Proposed Methodology

Eleftherios Lygouras[1], Ioannis M. Dokas[1(✉)],
Konstantions Andritsos[2], Konstantinos Tarchanidis[3],
and Antonios Gasteratos[1]

[1] Democritus Univeristy of Thrace, Xanthi, Greece
elygoura@ee.duth.gr, idokas@civil.duth.gr,
agaster@pme.duth.gr
[2] European University Cyprus, Nicosia, Cyprus
131536@students.euc.ac.cy
[3] Eastern Macedonia and Thrace Institute of Technology, Kavala, Greece
ktarch@teikav.edu.gr

Abstract. Search and rescue (SAR) teams are trained to swiftly act and operate in a feasible way, so as to reduce the risk of death associated with accidents due to natural and manmade disasters, as well as during recreational or sports activities such as swimming, climbing and hiking. In this context, SAR teams test emergent and rapidly evolving technologies to evaluate their usefulness in different operational scenarios and to assess if these technologies can increase their effectiveness. Unmanned Aerial Vehicles (UAVs) are examples of such technologies. UAVs are agile, fast and can exhibit autonomous behavior. Therefore, there is a need to assess their usefulness and learn their limitations aiming at comprehending under which settings and contexts can seamlessly collaborate with SAR teams. In our ongoing research, we have designed and implemented an autonomous UAV platform suitable for rescue and life-saving services, namely the ROLFER (Robotic Lifeguard For Emergency Rescue) platform. With a view to assess its operational behavior, as well as of other systems of similar characteristics, in cases where unwanted and adverse events occur during its operation, a new methodology is being proposed in this paper. The proposed methodology combines, in novel way, existing methods from different domains aiming at identifying the states and conditions under which the combined SAR and UAV system could fail. The methodology aims to support SAR teams in deciding if the UAV is appropriate tool to be used in the context of an emergency.

Keywords: Search and rescue missions · Drones · UAV · Simulation · Accident scenarios

1 Introduction

Unmanned Aerial Vehicles (UAVs) are rapidly emerging as useful tools for the execution of necessary tasks in many domains and are considered as one of the most versatile technologies of the future [1]. The potential benefits of UAV integration in

© Springer International Publishing AG 2017
I.M. Dokas et al. (Eds.): ISCRAM-med 2017, LNBIP 301, pp. 70–76, 2017.
DOI: 10.1007/978-3-319-67633-3_6

Search and Rescue (SAR) operations has been the spotlight in numerous research works [2, 3] with the objective to identify whether UAVs can perform their tasks effectively and enhance the performance of the teams in which they belong to, under different conditions.

Despite the immediate perceivable benefits of integrating UAVs into SAR operations, there are few barriers that "hinder" this intergradation. For instance, in the USA, legislatures are debating if and how UAV technology should be regulated, considering the benefits of their use, privacy concerns and their potential economic impact [4]. In addition, the legislation in some countries prohibits the use of UAV to collect information about or photographically or electronically record information about critical infrastructure without consent.

Considering this, someone could raise the following question. In case where an emergency occurs in a critical infrastructure and the SAR team uses a drone to collect electronic material, will this cause any problem or not? Apart for that there are some obvious questions that need to be answered. For instance, under which context the use of UAV could adversely affect, directly or indirectly, the mission of a SAR team? Alternatively, under which contexts and conditions the use of UAVs, instead of positively affect a SAR mission, will contribute to its failure?

If one considers, for instance, incidents like the one with the drone, which was flying few feet above a news helicopter that was covering a fire and that the U.S. Federal Aviation Administration receives 25 reports a month of drones flying too close to manned aircraft [5], then analogous questions could emerge in cases where UAVs are integrated into SAR missions. Ideally, these questions must be answered before deciding if it would be beneficial to integrate an UAV into a SAR mission. Currently, there are no research works that propose tools and methods based on which it would be possible to answer such questions in a scientific manner.

This early stage research work aims to describe a solution to this problem by introducing a novel methodology that combines methods and tools from safety and decision-making science and IT and simulation technologies. Consequently, the objective of this paper is to present for the first-time: (a) The selected methods and tools that comprise the proposed methodology and (b) How these tools are "interacting" in order to deliver what is required from the methodology.

It is assumed that the proposed methodology will: (a) Help decision makers and SAR team leaders to comprehend the levels of complexity of each mission where there is a possibility of UAV integration. (b) Provide tools to assess the risks of failures and accidents in the context of each mission and decide in favor or against the integration of the UAV into the mission. Our future goal is to translate these assumptions into research hypotheses and conduct research to answer them.

2   Needs

To achieve the aim of this research, it is necessary to combine into the proposed methodology different tools to address the following needs.

1. *UAV Specifications*: Obviously, a basic need is to identify the technical specifications of the UAVs i.e. its components, their interactions, their reliability and operational constraints. To satisfy this need one can review the technical documents from the manufacturers. Additionally, one can interview expert users and UAV pilots.

2. *Social System*: Typically, UAVs perform their tasks in coordination and collaboration with a team(s) of humans to achieve the goals in each SAR mission. In addition, there are humans or a team responsible for maintaining and piloting the UAV. In the context of this research there is a need to define the social system that collaborates with the UAV. There is a need to define the structure of the team the responsibilities of each team member, constraints that should be enforced under certain conditions. To satisfy this need one should establish contact with the SAR team and retrieve this information via manuals, official procedures and/or interviews.

3. *Mission Context*: Not all SAR missions are the same. Many things vary between SAR missions, even in missions with the same goal. Additionally, when a team executes the tasks in the same area for the same goal, the environmental conditions are not the same. In short, every mission is deployed in a different context. To satisfy this need one should collect data and information on the environmental conditions as well as of the terrain. He should also be aware of important information such as the position of obstacles and of other elements of the environment, which could interact with the UAV and with the SAR team. Thus, one should collect data such as, meteorological data or terrain data from maps or GIS applications and to survey the area of operations before the mission.

4. *Accident Scenarios*: Another basic need is the identification of scenarios leading to hazards and thus to losses, such as mission failure or injury or death, so they can be eliminated or controlled. To satisfy this need one should apply a hazard analysis method for a given mission. In addition, information and data collected to satisfy the previous needs most probably will be used during the application of the hazard analysis method to satisfy this basic need.

5. *Simulation*: Having the accident scenarios at hand, there is a need to simulate the behavior of the SAR – UAV team to "feel" or "experience" how problems could emerge. Specifically, one would need to identify how unwanted behaviors could emerge in the SAR – UAV team during the execution of the tasks in each mission and what could be the mechanisms that could trigger a loss. To satisfy this need one could utilize simulation tools for physics modelling, agent based and complex systems modelling, to name a few.

6. *Decision Making*: With a simulated experience of possible emergent problems in a mission, there is a need to decide about the integration of the UAV into the SAR team. To satisfy this need one could utilize expert judgments heuristics and tools based on multiple criteria decision making and naturalistic decision making.

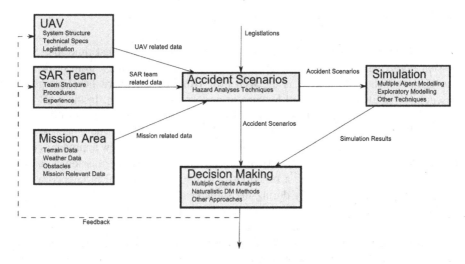

Fig. 1. The proposed methodological framework

3   A Methodological Framework

The proposed methodological framework is shown in Fig. 1. It depicts: (a) The actions necessary to address each need described above, (b) How the outputs from each action produce inputs to other actions to address other needs. It depicts, in short, how data, information, acts and needs are linked together.

The boxes in Fig. 1 represent the needs described in the previous section. Each box contains the objectives of the acts that are necessary to address each need. For example, to comprehend the need no 3 described in the previous section, about comprehending the mission context, it is necessary to acquire information and data in relation to the weather, the terrain, the potential obstacles to the SAR team and UAV and other relevant to the mission data.

The arrows going out from the boxes indicate the outputs produced by the acts which are necessary to satisfy each need. The direction of each arrow shows where these outputs will be used as inputs to address other needs. For example, the arrows going out from the boxes, "UAV", "SAR TEAM" and "MISSION AREA" express the UAV and SAR team related data as well as the mission related data which will be used as inputs, together with the legislations, for the execution of hazard analysis techniques, which are the necessary acts to address the need of the identification of "ACCIDENT SCENARIOS". The solid arrow going out from the "DECISION MAKING" box indicates the final decision in relation to the integration of UAVs with the SAR team in a specific SAR mission by the decision makes, after applying the proposed methodology. The dashed arrow indicates a feedback potential that could enforce changes to the SAR team and UAV system structure or to their integration procedures, after comparing the performance of the team and the outcome of the proposed framework in a specific SAR mission.

4   Towards Applying the Framework

The proposed methodological framework will be applied for the first time in a SAR mission involving the surveillance of large area to rescue swimmers from drowning. The UAV which will be used is the ROLFER (Robotic Lifeguard For Emergency Rescue). ROLFER does not need humans to perform its navigation. In the SAR scenario where the proposed methodology will be tested for the first time, ROLFER will provide one or more auto-inflatable floats to a distressed swimmer. The objective is to provide immediate help to the swimmer until the lifeguard rescue crew approaches him. The SAR team will be the lifeguard crew of the beach where the mission will take place. The mission area will be a beach close to the city of Xanthi in Greece where people are swimming especially during the summer months of the year.

Data about ROLFER's system architecture its elements and interactions have been collected. In addition, the designers of the ROLFER have agreed to provide additional data and information any time it will be required. The research team has established linkages with the coast guard crew responsible for the SAR missions in the study area. During the time of writing this paper, data and information is collected about the structure and procedures of the lifeguard crew. Data and information about the mission area, can be collected via maps, aerial photos, weather feeds from the Hellenic National Meteorological Service etc.

The collected data described above will be utilized in the near future as inputs to the Systems-Theoretic Process Analysis (STPA) hazard analysis [6]. STPA will be used to identify scenarios leading to identified hazards and thus to losses so they can be eliminated or controlled [7]. STPA will be used because its theoretical basis is based on systems theory, contrary to the traditional hazard analysis techniques which have reliability theory at their foundation. As result, STPA will help analysts identify how the integrated ROLFER and lifeguard crew system will violate its safety constraints, that is, why the appropriate constraints are not imposed on the interactions of the integrated system.

The accident scenarios generated by STPA will provide the inputs and the context for the simulations of the behavior of the integrated ROLFER and lifeguard crew system under different conditions. For the simulations, it is planned to use the Mission Planner [8], that is an open source software allowing the simulation and the full flight-control of multicopters, planes and rovers, where the drone flight can be programmed in Iron Python scripting.

Figure 2 depicts the user interface of the Mission Planner software. The interface in this image is divided in 3 compartments. The first compartment displays a map showing the area where the UAV will operate. Specifically, the map shows the: (a) Location of the base station of ROLFER, that is its main controlling computer (a Windows PC), an Android mobile phone or a tablet interconnected through USB connection, a USB telemetry link to communicate with the drone and the remote controller for the manual operation of the drone, in case it is apparent. (b) Exact position of ROLFER and its

Fig. 2. Aspect of the Mission Planner with the ROLFER path during a mission

current heading. (c) Location of the distressed swimmer. (d) Path which the ROLFER must follow to reach the distressed swimmer and return to its base station. The second compartment of the user interface contains the acetometer and horizon indicators and other flight data. The third compartment contains information about the current altitude of the UAV.

To the above tools we plan to integrate agent based modelling software, such as NetLogo [9], to include the behavior of moving elements, people and the lifeguard crew within the mission area.

Lastly, the simulation results will be used by the lifeguard crew to assess the risks in utilizing ROLFER in different contexts of the same mission and for the same mission area. During this phase, naturalistic decision-making approaches and/or multiple criteria analyses methods will be utilized to justify in a scientific manner the lifeguard crew decisions.

5   Concluding Remarks

Unquestionably UAVs can positively support many human activities. However, the integration of UAVs with teams towards achieving a purpose, such as a SAR mission, could have the opposite effect from the one that is desired under specific conditions. Consequently, there is a need to define risk assessment methodologies aiming at enhancing team-UAVs integration for specific contexts and for specific missions.

This paper introduced a work in progress that aims for a solution to this problem. Specifically, it introduced a novel methodological framework that combines methods and tools from safety and decision-making science and IT and simulation technologies with the objective to support decision makers to decide if the integration of a UAV with a SAR team will be beneficial or not in specific contexts.

The proposed methodology is currently at a testing phase. The goal is to apply the methodology in different SAR mission and in different contexts to assess if it can help decision makers and SAR team leaders to comprehend the levels of complexity of each mission where an UAV is planned to be used. An additional goal is to evaluate if the proposed approach will be considered of added value to be used by SAR teams in practice to assess the risk of SAR missions with UAV integration.

References

1. Mickel, C.: Lindabury, McCormick, Esabrook & Cooper, PC, Westfield, NJ: Despite Clear Benefits, the Construction Industry is Slow to Integrate Unmanned Aerial Vehicles into Projects, Under Construction, Vol. 17 No. 4 (2016)
2. De Cubber, G., Doroftei, D., Serrano, D., Chintamani, K., Sabino, R., Ourevitch, S.: The EU-ICARUS project: developing assistive robotic tools for search and rescue operations. In: IEEE International Symposium on Safety, Security, and Rescue Robotics (SSRR), pp. 1–4. IEEE Press, Piscataway (2013)
3. Waharte, S., Trigoni, N.: Supporting search and rescue operations with UAVs. In: Proceedings of the 2010 International Conference on Emerging Security Technologies, EST 2010, pp. 142–147. IEEE Computer Society, Washington, DC (2010)
4. NCSL Current Unmanned Aircraft State Law Landscape. http://legislature.vermont.gov/assets/Documents/2016/WorkGroups/House%20Judiciary/Bills/S.155/S.155 ~ Erik%20FitzPatrick ~ NCSL%20State%20and%20Federal%20Drone%20Law%20Summary ~ 3-15-2016.pdf
5. 12 drone disasters that show why the FAA hates drones. http://www.techrepublic.com/article/12-drone-disasters-that-show-why-the-faa-hates-drones/
6. Leveson, N.: Engineering a Safer World: Systems Thinking Applied to Safety. The MIT Press, Cambridge (2012)
7. An STPA Primer, Version 1, August 2013 (Updated June 2015). http://psas.scripts.mit.edu/home/wp-content/uploads/2015/06/STPA-Primer-v1.pdf
8. Mission Planner Home. http://ardupilot.org/planner/index.html
9. NetLogo Home Page. https://ccl.northwestern.edu/netlogo/

Towards Developing Dependable Systems Suitable for Crisis Management Applications

Angeliki Zacharaki[1]([✉]), Ioannis Kostavelis[2], and Ioannis Dokas[1]

[1] Department of Civil Engineering, University of Thrace,
Building A- Campus Democritus, 671-32 Xanthi, Greece
{azachara,idokas}@civil.duth.gr
[2] Department of Production and Management Engineering,
University of Thrace, Vas. Sophias 12, 671-32 Xanthi, Greece
gkostave@pme.duth.gr

Abstract. Response in crisis situations retains important risk mainly due to increased amount of unpredictable conditions at the incident scenery. Efficient management of such hazardous events necessitates the utilization of autonomous systems capable of operating in a dependable manner without placing the human life in danger. To achieve this, the design of rescue and exploration systems should retained increased operational capabilities, yet coupled with multilevel dependability. The paper at hand, aims to identify the different level of dependabilities that such a system should retain in order to operate in various crisis situations, while at the same time will provide conceptual solutions on how the different dependability levels can be endowed to the system. Moreover, it is proved that the endowed dependability should be maintained iteratively during the system's lifespan by exploiting systems' theory techniques preserving system's dependable behavior in crisis management situations.

Keywords: Dependability · Autonomous systems · System analysis · Crisis management

1 Introduction

In accordance to Laprie, J.C. [1], dependability is a system concept that integrates such attributes as reliability, availability, safety, confidentiality, integrity, and maintainability. However, the term dependability has been introduced in early 1980 to outline aspects like fault-tolerance and system reliability. Since then, all these attributes are considered mandatory for contemporary semi/unsupervised systems. Moreover, dependability defines the amount of autonomy a contemporary system retains [2]. Therefore, the need for highly dependable systems, which act autonomously, is more intense in crisis management situations, where increased chance of risk is present due to unpredictable conditions occurring at the scenery of the incident. The matters are even worse when it comes to hazardous events where it is likely to endanger the life of the rescuer workers, as well as any first responder. After the 9/11 World Trade Center collapse rescue

© Springer International Publishing AG 2017
I.M. Dokas et al. (Eds.): ISCRAM-med 2017, LNBIP 301, pp. 77–84, 2017.
DOI: 10.1007/978-3-319-67633-3_7

systems have been widely utilized while the main concern of the respective scientific community is the construction of dependable systems allowing autonomous operation without jeopardizing human lives [3].

Towards this direction, the main objective of this work is to outline a hierarchical framework for measuring and defining a system's dependability during its construction phase, so as to retain increased human trust in hazardous scenarios. For the design phase, hardware and software tools required to create a dependable system are defined, while for the runtime phase, system theory methodologies are discussed offering thus a complete set of tools for the preservation of systems' trust to operate in crisis situations. To the best of our knowledge this work comprises a first attempt to systematically introduce and outline the dependability measurements for systems in crisis situations.

2   Background on Contemporary Autonomous Systems

Albeit the fact that, great amount of research has already been conducted in the respective domain for the determination of aforementioned dependability attributes, only few works that measure and quantify these system abilities exist. The authors in [4] aim to formulate a generic mathematical model for the design of fault-tolerance systems in order to increase safety and reliability. However, a dependable system that operates in crisis events needs to be self-aware, retaining consciousness not only about its faults but also about the impact of these faults to its abilities to address the targeted missions. To achieve this, during system design, the mapping of system's functionalities into operational components is mandatory. Towards this direction, the authors in [5] reported the development of a system-specific tool based on Unified Modeling Language (UML) suitable to check faults in component based level. In a more contemporary work, the authors in [6] developed a system-independent collaborative tool suitable for the architecture design of robotic systems. This tool followed a hierarchical structure ensuring a complete mapping among high level functionalities to low level system specifications ensuring local dependability of each subordinate module of the system. However, this tool was oriented only towards the development view point of the system design and not to its physical implementation In order to decrease the risk about the civilians exposed in hazardous events during crisis situations, protection agencies around the world emerged the employment of robotics systems [7]. Towards addressing this recommendation, automation and robotics scientific laboratories steered their attention to the development of robotic systems targeting operation in crisis situation such as urban search and rescue (USAR) [8], physical disasters [9] and terrorism [10]. More precisely the authors, in [11] developed the Hybrid Exploration Robot for Air and Land Deployment (HER-ALD) which was three-in-one autonomous system able to operate in air and land while retained snake-like movement capabilities. It is evident that such a complex and hyper-terrain operation system is challenging to meet adequate dependability and, thus autonomy levels. This can be proved by the fact that it was able to operate for solely 17 min with maximum payload something that comprises a limitation during cooperation with rescue workers. Murphy, R. in her book namely, "Disaster Robotics" [3] introduced that contemporary robots designed for crisis situations should retain advanced cognitive, perception, manipulation capabilities, all coupled with time endurance and reliability which is the primary component of dependability. Considering the domain of counter-terrorism with robotics applications several solutions have been developed during the last years, the most established of which is the collaboration with explosive ordnance disposal (EOD) teams aiming to neutralize explosive mechanisms. Such robots have been widely utilized as remote access and manipulation means from specialized teams on explosive counteraction [12]. Albeit the fact that such robots have been successfully utilized in real missions, their autonomy was limited since in their majority they were tele-operated by humans. It is evident that robots designed to operate in crisis situations exhibit limited dependability and, this stems from the deteriorated trust the human rescuers can put on their robotic co-workers to undertake and accomplish a target mission.

3 Definition of Systems Dependability and Measures to Increase Trust in Crisis Applications

Considering the amount of research that has already been conducted in the respective domain, a modern definition of the system dependability can be provided such as: *dependability outlines the ability of the system to perform its designated tasks without systematic error specifying at the same time the amount of trust that humans can put on the system to operate without their supervision.* Taking for example the RAMCIP robot [13] which aims to assist elderly people with early Alzheimer, the dependability of this system is expressed as the trust we put on it to look after a patient for one day. Herein, six levels of dependability are exhibited along with their respective techniques to meet each level's requirements. Level-0 to Level-2 is oriented towards the techniques existing that ensure hardware related dependable behavior, while Level-3 to Level-5 is related to the strategies should be followed in software level that endow autonomy to the designed system. Moreover, the identified levels should be addressed during the design phase in order the system to retain the respective dependable behavior during operational phase.

Level-0: Zero Dependability
This level is appended for symbolic purposes since all the systems have a certain degree of dependability, even the preliminary prototypes.

Addressing Level-0:
This basic level of dependability can be achieved by identifying the user's need for the construction of system with specific functionalities. Then, the design of the target application considering fundamental requirements and specifications can meet Level-0 of system dependability.

Level-1: Mean Time Between Failure
In a system targeted to operate in crisis management situations the exact knowledge of the mean time between failure (MTBF) for each subordinate component

is imperative. This knowledge may prevent the deployment of a system in a hazardous event when it is close to a failure period.

Addressing Level-1:
During the development of autonomous systems -and especially robotic applications- coupling of hardware components with a well established communication framework can eliminate discontinuities and faults. This characteristic is provided during the early development of each component by establishing mechanisms to preserve *operation log data*. This attribute can provide the means for the statical analysis on MTBF for the majority of the hardware components both during runtime and a posteriori facilitating thus postmortem analysis.

Level-2: Safe failure of the system
The system should be designed in such a manner that fail-safe mechanisms will live into it. This will enable it to halt its and place it into a safe mode when failures are detected.

Addressing Level-2:
This can be achieved by considering criticality ranking and prioritization among the operation components of the system counting in the significance of its failure, the time scale and the risk associated with its tasks. Embedding of self diagnostics of its component in the hardware level provides such attributes, where the system will have the ability to interpret the diagnostics and re-initialize part of these components. Criticality ranking of the components can be also identified during the architecture design by utilizing specialized tools such as ArchGenTool [6] which can host characteristics such as the level of impact for each component.

Level-3: Recovery from failures and gentle degradation
The system should be able to recover from a portion of failures by restarting its operation, while it should recognize the impact of its failure on its functions being able compensate on its own.

Addressing Level-3:
This level of dependability should act as the middleware interphase among the hardware and software components. This middleware should be responsible to gather the measurements obtained from each side and by assessing their values to identify if a sub-system is critical or about to fail. By excluding this sub-system from the chain the system should degrade in controlled time and not during the operational phase. One of the most well established middlewares is the Robot Operating System (ROS) [14] which is a platform that offers robust interfacing among hardware and low-level control software. ROS has the capacity to synchronize un-synchronous sensors while at the same time offers significant attributes for Level-1 to Level-2 dependability.

Level-4: Task based dependability
The system should be able to recognize the impact of a failure on the overall task it is undertaking and re-task activities in order to minimize the impact of the failure to the overall task.

Addressing Level-4:
This level of dependability is related more on the software architecture of the designed system. Thus, ultra-modular architecture should be established during the design phase. Moreover, each module or software component should retain self diagnostics, metrics and error-handling so as to propagate its unavailability to other software components. This will prevent task failures during crisis management events.

Level-5: Mission based dependability
The system should be able to recognize the impact of a failure on the overall objectives of a mission and communicate the nature of the failure to other systems or humans so as to minimize the impact on the mission objectives. It is common that autonomous systems work together with human rescue teams in crisis events and, therefore, these systems should have the ability to identify their weaknesses towards subordinate tasks of the missions and communicate them to the collaborative teams ensuring awareness and mission continuation.

Addressing Level-5:
This is the highest level of dependability of a system and refers mainly to the high-level software components that endow advanced cognition, perception as well as decision making capabilities. These components should be designed in a fashion to act like the conductors for their decencies changing their modes depending on the system defined tasks. Recording and self awareness of this highly sophisticated software components will ensure re-ordering of the tasks to deal with a sub-system failure.

It is evident that Level-4 and Level-5 system's dependability depends on the methodology followed for the development of the software components of each system. Modular software implementations coupled with error handling in hardware level are the fundamentals for designing of autonomous systems capable to offer mission-based dependability.

4   Hazard Analysis Techniques on Dependable Systems

Although, dependability is an attribute that can be embedded to a system during its designing phase, preservation and supervision of system's dependability is also ensured during its entire lifespan. This can be achieved by iteratively assessing its behavior during runtime, definition of long term interactions and relationships among components and modeling of complex factors that may lead to a system failure. Moreover, hazardous events modeling can reveal weakness of the design phase which necessitates the re-design of several functional components.

The authors in [15] introduced a Systems-Theoretic Accident Model and Processes (STAMP) which it based on systems theory and extends the traditional analytic reduction and reliability theories. It mainly advocates that accidents involve a complex dynamic process, so they are not simply chains of events and component failures. For this reason, STAMP theory views safety as an emergent property that arises when system components interact with each other within their larger environment. In a more contemporary work, the author in [16] outlined that Systems-Theoretic Process Analysis (STPA) is a hazard analysis technique that encapsulates the principles of the STAMP model. STPA also identifies causal factors not fully handled by traditional hazard analysis methods, such as software errors, component interactions, decision-making flaws, inadequate coordination and conflicts among multiple controllers, and poor management and regulatory decision-making. For example, the authors in [17] applied take-off hazard analysis on an Unmanned Aerial Vehicle (UAV) taking advantage of STPA method to address system safety as a control problem.

Moreover, the Early Warning Sign Analysis (EWaSAP) is an add-on tool to STPA [16,18] and its aim is to provide a structured method for the identification of early warning signs required to update mental models of system agents. Under this approach, EWaSAP introduces an additional type of control action, the awareness action. An awareness control action is required from a controller who must provide warning messages and alerts to other controllers inside or outside the system boundaries whenever data indicating the presence of threats or vulnerabilities is perceived and comprehended [18]. According to [19] the Risk SituatiOn Awareness Provision (RiskSOAP) indicator used to measure the capability of a complex sociotechnical system to provide its agents with situation awareness about the presence of its threats and vulnerabilities. The methodology uniquely combines three methodos: (1) the STPA hazard analysis, (2) the EWaSAP early warning sign identification approach and (3) a dissimilarity measure for calculating the distance between binary sets. Towards this direction, the early work presented by in [20] demonstrated the application of RiskSOAP to a robotic system and the findings show that the indicator is an objective measure for the system's capability to provide its agents with situation awareness about its threats and vulnerabilities.

5   Discussion and Future Work

In the present theoretical work, the necessity for the construction of dependable systems capable of operating autonomously in hazardous situations has been discussed. The fundamental levels required for a system's dependable behavior have been identified, while appropriate strategies towards addressing of each level of dependability during the design phase have been presented. The levels of dependability are organized in a hierarchical fashion indicating that each one comprises the prerequisite to ensure the next level of dependability. Lower dependability levels ensure less trust that the humans can put on the system to operate autonomously and, thus, only simple missions can be addressed. As we ascending in the hierarchy of the dependability level, the level of trust we can put on the system to operate increases, i.e. resolve a crisis mission. However, the herein identified tools and means to reach a desired dependability level is only a fraction of the existing solutions. Continuous dependability monitoring has been proved as a valuable tool for crisis management with autonomous applications. In our future work we intend to adapt the RiskSOAP methodology to the specificity of crisis management while designing a system with dependability of level-5. This will ensure the evaluation of the above mentioned theoretical work on a real system comprising robots operating in crisis situations.

References

1. Laprie, J.C.: Dependability: Basic Concepts and Terminology: In English, French, German, Italian and Japanese, vol. 5. Springer, Wien (2013). doi:10.1007/978-3-7091-9170-5

2. Dubrova, E.: Fundamentals of dependability. Fault-Tolerant Design, pp. 5–20. Springer, New York (2013). doi:10.1007/978-1-4614-2113-9_2

3. Murphy, R.R., Tadokoro, S., Kleiner, A.: Disaster robotics. In: Siciliano, B., Khatib, O. (eds.) Springer Handbook of Robotics, pp. 1577–1604. Springer, Cham (2016). doi:10.1007/978-3-319-32552-1_60

4. Buckl, C., Knoll, A., Schieferdecker, I., Zander, J.: 10 Model-Based Analysis and Development of Dependable Systems. In: Giese, H., Karsai, G., Lee, E., Rumpe, B., Schätz, B. (eds.) MBEERTS 2007. LNCS, vol. 6100, pp. 271–293. Springer, Heidelberg (2010). doi:10.1007/978-3-642-16277-0_10

5. Jürjens, J., Wagner, S.: Component-based development of dependable systems with UML (2005)

6. Ruffaldi, E., Kostavelis, I., Giakoumis, D., Tzovaras, D.: ArchGenTool: a system-independent collaborative tool for robotic architecture design. In: International Conference on Methods and Models in Automation and Robotics, pp. 7–12. IEEE (2016)

7. Fink, S.: Crisis Management: Planning for the Inevitable. American Management Association, New York (1986)

8. Looije, R., Neerincx, M., Kruijff, G.J.M.: Affective collaborative robots for safety & crisis management in the field. In: Intelligent Human Computer Systems for Crisis Response and Management, (ISCRAM 2007), Delft, Netherlands, vol. 1 (2007)

9. Zarzhitsky, D., Spears, D.F., Spears, W.M.: Distributed robotics approach to chemical plume tracing. In: International Conference on Intelligent Robots and Systems, pp. 4034–4039. IEEE (2005)

10. Murphy, R.R., Kleiner, A.: A community-driven roadmap for the adoption of safety security and rescue robots. In: International Symposium on Safety, Security, and Rescue Robotics, pp. 1–5. IEEE (2013)

11. Latscha, S., Kofron, M., Stroffolino, A., Davis, L., Merritt, G., Piccoli, M., Yim, M.: Design of a hybrid exploration robot for air and land deployment (herald) for urban search and rescue applications. In: International Conference on Intelligent Robots and Systems, pp. 1868–1873. IEEE (2014)

12. Danna, V.: A new generation of military robots. RESEARCH Appl. AI Comput. Animation **7**, 2–5 (2004)

13. Kostavelis, I., Giakoumis, D., Malasiotis, S., Tzovaras, D.: RAMCIP: towards a robotic assistant to support elderly with mild cognitive impairments at home. In: Serino, S., Matic, A., Giakoumis, D., Lopez, G., Cipresso, P. (eds.) Mind-Care 2015. CCIS, vol. 604, pp. 186–195. Springer, Cham (2016). doi:10.1007/978-3-319-32270-4_19

14. Quigley, M., Conley, K., Gerkey, B., Faust, J., Foote, T., Leibs, J., Wheeler, R., Ng, A.Y.: ROS: an open-source robot operating system. In: ICRA Workshop. vol. 3, p. 5. Kobe (2009)

15. Leveson, N.: Engineering a Safer World: Systems Thinking Applied to Safety. MIT press, Cambridge (2011)
16. Leveson, N.: A systems approach to risk management through leading safety indicators. Reliab. Eng. Syst. Safety **136**, 17–34 (2015)
17. Chen, J., Zhang, S., Lu, Y., Tang, P.: STPA-based hazard analysis of a complex UAV system in take-off. In: International Conference on Transportation Information and Safety, pp. 774–779. IEEE (2015)
18. Dokas, I.M., Feehan, J., Imran, S.: Ewasap: An early warning sign identification approach based on a systemic hazard analysis. Saf. Sci. **58**, 11–26 (2013)
19. Chatzimichailidou, M.M., Dokas, I.M.: Risksoap: introducing and applying a methodology of risk self-awareness in road tunnel safety. Accid. Anal. Prev. **90**, 118–127 (2016)
20. Chatzimichailidou, M.M., Dokas, I.M.: Introducing risksoap to communicate the distributed situation awareness of a system about safety issues: an application to a robotic system. Ergonomics **59**(3), 409–422 (2016)

Decision Making in the Context of Crisis Management

Hybrid Soft Computing Analytics of Cardiorespiratory Morbidity and Mortality Risk Due to Air Pollution

Vardis-Dimitris Anezakis[1(✉)], Lazaros Iliadis[2],
Konstantinos Demertzis[1], and Georgios Mallinis[1]

[1] Department of Forestry & Management of the Environment
& Natural Resources, Democritus University of Thrace,
193 Pandazidou St., 68200 N Orestiada, Greece
{danezaki,kdemertz,gmallin}@fmenr.duth.gr
[2] School of Engineering, Department of Civil Engineering,
Democritus University of Thrace,
University Campus, Kimmeria, 67100 Xanthi, Greece
liliadis@civil.duth.gr

Abstract. During the last decades, climate change has been contributing significantly to the increase of Ozone and Particulate Matter in major urban centers. This might result in additional enhancement of serious seasonal respiratory and cardiovascular diseases incidents. This research effort introduces an innovative hybrid approach that fuzzifies the involved features. The final target is the development of a Mamdani fuzzy inference system with weighted fuzzy rules. The system's output comprises of the partial meteorological and air pollution risk indices per season. Fuzzy conjunction T-Norms have been employed to estimate the unified risk index. Moreover, the effect of one to seven days delay regarding high values of the above indices to the morbidity and mortality indicators in the prefecture of Thessaloniki has been studied. Hybrid Fuzzy Chi Square Test has been performed to identify the degree of dependences between the unified air pollution-meteorological risk indices and serious health even mortality cardiorespiratory problems.

Keywords: Mamdani fuzzy inference system · Fuzzy chi square test · Fuzzy risk indices · Air pollution index · Mortality · Morbidity · Thessaloniki

1 Introduction

Air pollution (AIPO) is one of the most serious problems of modern societies. It is related to the concentration of all sorts of substances, to radiation or other forms of energy in quantity or duration, which can cause an increase in hospital admissions and a reduction in life expectancy. Under certain conditions, concentrations of primary (CO, NO, NO_2, SO_2) or secondary pollutants (O_3) are likely to reach critical alarm levels that create inappropriate living conditions, increasing the incidence of myocardial infarction and stroke. The quantification of urban residents' morbidity and mortality due to meteorological conditions that intensify concentrations of air pollutants is considered a

© Springer International Publishing AG 2017
I.M. Dokas et al. (Eds.): ISCRAM-med 2017, LNBIP 301, pp. 87–105, 2017.
DOI: 10.1007/978-3-319-67633-3_8

multifactorial and complex problem. It is usually necessary to carry out an in-depth spatio-temporal analysis of extreme meteorological conditions that maximize the problem as well as to discover the temporal delay of the atmospheric conditions' influence on the morbidity and mortality of urban population. Another key issue is the discovery of the combined effect of meteorological conditions and pollutants in urban centers.

This paper introduces an innovative air pollution analytics approach with its corresponding system. One of its basic contributions is the creation of fuzzy weighted rules applied on the parameters that determine the morbidity and mortality of the inhabitants of Thessaloniki city, on a seasonal basis. Two partial Risk indices (produced by fuzzy rules) have been developed to analyze the daily Risk due to meteorological parameters (DRMP) and due to air pollutants respectively (DRAP), on a seasonal basis. In addition, a unified Risk index (URI) has been introduced by applying fuzzy T-Norm conjunction operation on the DRMP and DRAP partial indices.

Correlation analysis has been performed between the *Linguistics/Fuzzy sets (L/FS)* corresponding to the above three developed indices and the morbidity and mortality Risk *Linguistics* in Thessaloniki using the hybrid Fuzzy Chi Square Test approach.

In general, the use of this hybrid approach determines not only the existence of dependences but also the degree of dependence among the *L/FS*. This is achieved by fuzzyfying the Chi-Square Test P-Values. This dependence is expressed by using properly designed L/FS such as *Low*, *Medium* and *High* obtained by fuzzy membership functions (FMF).

1.1   Related Literature - Motivation

Various studies [2, 19, 23] in the literature have analyzed the combined effect of meteorological parameters, atmospheric pollution and socio-economic characteristics of residents on morbidity and mortality from cardiovascular and respiratory diseases.

Sarigiannis et al. [17] proposed a survey, which deals with the assessment of health impact and the respective economic cost attributed to particulate matter (PM) emitted into the atmosphere from biomass burning for space heating, focusing on the differences between the warm and cold seasons in 2011–2012 and 2012–2013 in Thessaloniki. Vlachokostas et al. [21] presented a methodological approach in order to estimate health damages from particulate and photochemical urban air pollution and assess the order of magnitude as regards the corresponding social costs in urban scale. Zoumakis et al. [24] analyzed mortality during heat waves in Northern Greece, separately for June, July and August, during the period 1970–2009, and to quantify a preliminary relationship between heat stress and excess mortality, taking into consideration the population adaptation to the local climate. Kassomenos et al. [10] studied the sources and the factors affecting the particulate pollution in Thessaloniki. Hourly PM_X concentrations from two monitoring sites were therefore correlated to gaseous pollutant concentrations and meteorological parameters during the 2-year period between June 2006 and May 2008. Papanastasiou et al. [15] used meteorological and AIPO data which observed in Athens, Thessaloniki and Volos were analyzed to assess the air quality and the thermal comfort conditions and studied their synergy, when extreme hot weather prevailed in Greece during the period 2001–2010. Kyriakidis et al. [12] employed a number of Computational Intelligence algorithms to study the forecasting of the hourly and daily Common Air Quality Index. These algorithms included artificial neural networks, decision trees and regression models combined with different datasets. Voukantsis et al. [22] formulated and employed a novel hybrid scheme in the selection process of input variables for the forecasting models, involving a combination of linear regression and artificial neural networks (multi-layer perceptron) models. The latter ones were used for the forecasting of the daily mean concentrations of PM_{10} and $PM_{2.5}$ for the next day. Prasad et al. [16] developed an ANFIS model predictor which considers the value of meteorological factors and previous day's pollutant concentration in different combinations as the inputs to predict the 1-day advance and same day air pollution concentration. The concentrations of five air pollutants and seven meteorological parameters in the Howrah city during the period 2009 to 2011 were used for development of an ANFIS model.

This research paper has been motivated by the lack of an integrated approach that would estimate the cumulative effect of the atmospheric conditions' correlations over public health (viewing the problem in a holistic manner). This means that we consider the combinatorial effect of meteorological features and air pollutants' concentrations. Additionally, to the best of our knowledge there is no approach in the literature trying to model time-lag effects of atmospheric conditions on public health using soft computing techniques. Specifically, there is lack of a holistic model that would analyze and realize the combinatorial effect of the meteorological-pollutants features to the respiratory and cardiovascular diseases incidents. On the other hand, the employment of the fuzzy X^2 test in environmental or medical cases is quite innovative.

Therefore, the proposed research attempts to fill a significant gap in the literature using fuzzy logic methods. It employs seasonal indices related to the parameters that maximize their values due to the processes and the mechanisms of the atmosphere, negatively affecting the public health of the inhabitants of Thessaloniki.

The basic innovation of the proposed model is based on the creation of the two partial risk indices the DRMP and the DRAP that were mentioned before. This is achieved through the employment of fuzzy weighted rules on a seasonal approach. More specifically for each season the system considers the meteorological conditions that maximize the problem of AIPO. In this way, it is possible to find the conditions for creating either the smoke cloud or the photochemical.

The determination of the unified risk index per season is based on the combined effect of the overall meteorological and the overall pollution indices, through the fuzzy conjunction T-NORM operators. The aim of the research is to find the dependences between the fuzzy linguistics corresponding to the three above indices with the morbidity and mortality L/FS not only on the same day but also with a delay of up to 7 days (0–7 days).

The produced system can be used by civil protection authority for the benefit of the residents of Thessaloniki. More specifically it can be used as a forecasting system on a short-term basis aiming to inform local authorities, citizens and hospitals on its output. This can significantly help the work of civil protection authority by providing alerts for hospitals due to specific pollution – weather conditions. In this way the hospital infrastructure will be ready to accept emergency incidents. This in-depth study strengthens civil protection mechanisms by serving as a tool for informing the public or hospitals on the potential of adverse meteorological conditions and smog or photo-chemical cloud development.

Additionally, finding correlations due to the delay of the effects of atmospheric conditions on public health can act as an information system for managing extraor-dinary and increased imports into public hospitals even several days after the occur-rence of adverse atmospheric conditions.

1.2 Research Data

Aiming to produce a symbolic representation of the complex correlations between atmospheric pollutants with morbidity and mortality in Thessaloniki, a statistical analysis of twelve AIPO measuring stations for the period 2000–2013 was carried out. The averages of the pollutants' values from the twelve stations were calculated as the selected stations record the atmospheric pollution from all the prefecture of Thessa-loniki. In most of the data records missing values were observed for periods of hours, days even months for the whole period of 2000–2013, probably due to malfunction. The SO_2, CO and O_3 pollutants are recorded hourly whereas only average daily values are measured for the PM_{10} and $PM_{2.5}$ particles.

The average daily values of CO and O_3 have been estimated after the calculation of the maximum daily average rolling values per 8 h.

Generally, all values are measured in $\mu g/m^3$, except for CO which is measured in mg/m^3. Except from the five air pollutants three meteorological features were employed namely: Air Temperature (AirTemp), Relative Humidity (RH), and sunshine hours (SUN) gathered from Thessaloniki airport station (Micra).

The following table contains a brief presentation of the measuring stations with statistical analytics (Table 1).

Table 1. Statistics of the measuring stations

ID	Station's name	Code	Missing values	Correct Data Vectors	Station's data
1	Aristotle University	APT	33.24%	12,679	O_3, NO, NO_2, SO_2
2	Agias Sofias	AGS	35.86%	18,272	O_3, NO, NO_2, CO, SO_2, PM_{10}
3	Kalamaria	KAL	54.43%	11,982	O_3, NO, NO_2, CO, SO_2, PM_{10}
4	Kordelio	KOD	31.83%	19,418	O_3, NO, NO_2, CO, SO_2, PM_{10}
5	Neochorouda	NEO	33.9%	9,656	O_3, NO, NO_2, PM_{10}
6	Panorama	PAO	32.18%	12,880	O_3, NO, NO_2, PM_{10}
7	Sindos	SIN	29.09%	20,199	O_3, NO, NO_2, CO, SO_2, PM_{10}
8	Egnatia	EGN	8.498%	31,419	O_3, NO, NO_2, CO, SO_2, PM_{10}, $PM_{2.5}$
9	25th March	25 M	5.68%	28,941	O_3, NO, NO_2, CO, SO_2, PM_{10}
10	Lagkadas	LAG	13.32%	26,594	O_3, NO, NO_2, CO, SO_2, PM_{10}
11	Eptapyrgio	EPT	12.01%	26,997	O_3, NO, NO_2, CO, SO_2, PM_{10}, $PM_{2.5}$
12	Malakopi		38.74%	18,797	O_3, NO, NO_2, CO, SO_2, PM_{10}
13	Macedonia Airport	Mikra	0.39%	40,751	Meteorological Station
14	AVERAGE OF ALL POLLUTION STATIONS		25.32%	21,429	O_3, NO, NO_2, CO, SO_2, PM_{10}, $PM_{2.5}$

During data pre-processing up to 75% of the daily hourly values or 75% of the hourly current 8-hour average values for CO and O3 were considered as coming from reliable measurements. As far as extreme prices are concerned, they were considered important as based on them emergency patient imports are intensified, thus it was considered appropriate not to remove them from our data.

In terms of disease selection, circulatory system diseases (I00–I99) and respiratory diseases (J00–J99) were categorized based on the tenth review of the ICD-10 International Statistical Classification of Diseases and Related Health Problems [9]. The rejection of patients' cases with injuries, traffic, poisoning, in-hospital illnesses or transported from other regional hospitals was necessary for the reliability of the data.

The collection of morbidity data was carried out by all public hospitals in the Prefecture of Thessaloniki. At county level, the collection of morbidity data for circulatory and respiratory diseases was carried out by the Hellenic Statistical Authority (ELSTAT).

2   Theoretical Frameworks and Methodology

2.1   Mamdani Fuzzy Inference System and Fuzzy Rules

Fuzzy logic is a modeling attempt close to the human way of thinking and inference, providing approximate reasoning mechanisms and inference/decision-making since the human brain tends to make the approximate reasoning based on qualitative perception criteria despite precise reasoning based a multitude of data.

In Mamdani FIS, human knowledge is illustrated in the form of fuzzy IF/THEN rules. The FRs are a mechanism of knowledge representation which is reflected in the hypothetical proposals, peculiar to the human way of thinking. The fuzzy sets expressing Linguistics are combined together and create Mamdani FR representing the knowledge available about a system. They are expressed in the following form:

$$IF\ x\ is\ A\ (Linguistic\ antecedent)\ THEN\ y\ is\ B\ (Linguistic\ consequent) \qquad (1)$$

where A and B are L/FS corresponding to parameters x and y which are defined in the universe of discourse X and Y respectively. The expression "x is A and y is B" are fuzzy proposals. The above rule 1, defines a fuzzy implication relation (between the parameters x and y) which correlates the truth level of the antecedent to the corresponding truth of the consequent. It is a fact that x belongs to the L/FS A with a degree of membership (DOM) and the same stands between y and B. Finally the output of the consequent is de-fuzzified by a proper function (centroid) and we obtain a crisp value which is a final numeric result. Its benefit is that it can be handled by a computational approach or a sensor [8].

The implementation of the Mamdani Inference System containing the corresponding FRs was done under the Matlab platform that offers an integrated fuzzy development, analysis and visualization environment [18].

2.2   T-Norm Fuzzy Conjunction

This paper attempts to calculate the URI, resulting from the cumulative effect of all the related factors, after performing integration operations on all individual fuzzy sets. This task is carried out, by using specific fuzzy conjunction "AND" operators known as T-Norms in the literature. The *Min*, the *Algebraic*, the *Drastic*, the *Einstein* and the *Hamacher Products* act as *T-Norms* [3, 5, 8, 11, 13, 14]. The *Min T-Norms* are unifiers of partial risk indices and they are quite optimistic as they are assigning the minimum risk value to the overall index [7, 8].

2.3   Chi-Square Test and Fuzzy Chi Square Test

The Chi-Squared hypothesis-testing is a non-parametric statistical test in which the sampling distribution of the test statistic is a chi-square distribution when the null hypothesis is true. The null hypothesis H_0 usually refers to a general statement or default position that there is no relationship between two measured phenomena, or no difference among groups. The H_0 is assumed to be true until evidence suggest otherwise [4, 6]. The statistical control index used for this assessment is the test statistic X^2.

$$X^2 = \sum \frac{(f_o - f_e)^2}{f_e} \tag{2}$$

where f_e is the expected frequency and f_o the observed one. The degrees of freedom are estimated as in Eq. 3 (for the table of labeled categories with dimensions *rXc*) (Tables 2 and 3).

$$df = (r - 1)(c - 1) \tag{3}$$

For the H_0 the critical values for the test statistic are estimated by the X^2 distribution after considering the degrees of freedom. If the result of the test statistic is less than the value of the Chi-Square distribution then we accept H_0 otherwise we reject it [6].

The calculated p-value prices include the possibility of error in the range [0–1]. Each error value is multiplied by 10, raised to the negative sixth power (p-value). The p-value equal to a is considered as boundary and it cannot determine the dependence or independence between the variables. The dependence is defined with p-values < a, whereas the independence with p-values > a.

Subsequently, p-values are fuzzified by using *Fuzzy Chi-Square test*, according to the specified confidence interval and to the significance level. This process of course requires the use of proper FMF, developed to estimate the level of dependence when *p-value < a* (in the closed interval [0-0.049999]) and the level of independence when p-value > a (in the closed interval [0.050001-1]). This is done to produce the proper *Linguistics*.

Table 2. Indicative dependence DOM using Fuzzy Linguistics (Interval [0-0.049999])

P-Value	Linguistics	High	Medium	Low
0	High	1	0	0
0.00001	High	0.99945	0	0
0.012499	High	0.37505	0.37495	0
0.0125	High/Medium	0.375	0.375	0
0.012501	Medium	0.37495	0.37505	0
0.03	Medium	0	0.75	0
0.035	Medium	0	0.5	0.2500125
0.037499	Medium	0	0.37505	0.374968
0.0375	Low	0	0.375	0.3750187
0.049999	Low	0	0	1

Table 3. Indicative Independence DOM using Fuzzy Linguistics (Interval [0.050001-1]

P-Value	Linguistics	Low	Medium	High
0.050001	Low	1	0	0
0.05001	Low	0.99997	0	0
0.287550	Low	0.374871	0.374868	0
0.287551	Medium	0.374868	0.374871	0
0.4301	Medium	0	0.75	0
0.71505	Medium	0	0.5	0.25013157
0.762518	Medium	0	0.375051	0.375047
0.762519	High	0	0.375048	0.375050
1	High	0	0	1

The following MATLAB commands were used to enhance the above for the Linguistics *Low, Medium and High.*

$$HighDependence = trimf(dependence, [-0.020000\ 0.000000\ 0.020000])$$
$$MediumDependence = trimf(dependence, [0.005000\ 0.025000\ 0.045000])$$
$$LowDependence = trimf(dependence, [0.030000\ 0.049999\ 0.070000])$$

$$LowIndependence = trimf(independence, [-0.329900\ 0.050001\ 0.430001])$$
$$MediumIndependence = trimf(independence, [0.145100\ 0.525100\ 0.905000])$$
$$HighIndependence = trimf(independence, [0.620000\ 1.000000\ 1.380000])$$

3 Description of the Proposed Methodology

The basic proposed modeling methodology includes the fuzzification of the crisp values of all considered parameters with triangular membership functions and their categorization by using risk Linguistics. In addition, fuzzy weighted rules were developed to produce Meteorological and Atmospheric Pollution risk indices respectively. Five T-Norms were used to perform Conjunction and create a unified index from the combination of the meteorological and the pollution indices.

A hybrid approach based mainly on the Fuzzy Chi-Square Test was applied to find the degree of dependences not only on the same day but also with a latency of up to 7 days between the Linguistics related to the three indices (*DRMP, DRAP, URI*) and the morbidity (MBI) and mortality (MTI) indices of the underlying diseases. Total research is interpreted in four distinct stages, which are analyzed below:

Step1. For each season of the year, meteorological conditions that maximize the problem of air pollution were investigated. By studying the literature related to winter [1, 20], the values of the minimum temperature, maximum humidity and sunshine were fuzzified. Based on the created fuzzy sets, the *Meteorological Smog Index (MSMI)* was defined and used. The combined effect of the lower minimum temperature values with the higher maximum humidity and the lower values of the sunshine hours corresponded to the Linguistic *High* with an assigned weight equal to 1.

Similarly, for the summer months, fuzzy weighted rules were designed by fuzzifying the values of the maximum temperature, the minimum humidity and sunshine. The result was the determination of the *Meteorological Photochemical Cloud Index (MPCI)*. In the summer months, the combined effect of the higher maximum temperature values, of the lower minimum humidity values and the higher hours of sunshine corresponded to the Linguistic *High* with an assigned weight equal to 1.

On the other hand regarding spring and autumn we have examined the conditions created by both meteorological indices, as they are transitional seasons that sometimes contribute to the creation of the *smog cloud* and sometimes they favor the development of *photochemical cloud*. We have examined only the pollutants CO, SO_2, PM_{10}, $PM_{2.5}$ for the days that MSMI was assigned a risk Linguistic higher than the one of MPCI.

The crisp values of the features CO, SO_2, PM_{10}, $PM_{2.5}$ were fuzzified to Risk Linguistics, whereas the combination of all fuzzy sets led to the definition of the *Pollution Index due to Smog (PIDS)*. For the days that had a higher MPCI we have examined the O_3 that is the result of photochemical reactions. The crisp values of O_3 were fuzzified and thus the *Pollution Index due to Photochemical Cloud (PIDP)* was obtained. Regarding the days that both meteorological risk indices were assigned the same linguistic we have selected the index with the higher fuzzy DOM (Tables 4 and 5).

Step2. For all seasons, we have examined the effect of the temporal delay of the MSMI and the MPCI indices, to the morbidity MBI and mortality MTI indices for the prefecture of Thessaloniki.

Step3. Five Conjunction T-Norms (*Min, Algebraic, Einstein, Hamacher, Lucasiewicz*) [8] were used for all seasons, to develop a unified risk index from the fuzzy conjunction between the meteorological and the pollution indices, on a daily basis (MET_POL).

Table 4. Fuzzy Sets and the corresponding Linguistics of each feature

Parameters	L/FS (Linguistics used)
Air temperature- Relative Humidity	Very Low(VL), Low(L), Medium(M), High(H),Very High(VH)
Sunshine	Low(L), Medium(M), High(H)
MSMI	Low(L), Medium(M), High(H)
MPCI	Low(L), Medium(M), High(H)
PIDS	Low(L), Medium(M), High(H)
PIDP	Low(L), Medium(M), High(H)
RECAH	Low(L), Medium(M), High(H)
DDRE	Low(L), Medium(M), High(H)

*Meteorological Smog Index (MSMI)
Meteorological Photochemical Cloud Index (MPCI)
Pollution Index due to Smog (PIDS)
Pollution Index due to Photochemical Clous (PIDP)
Respiratory Cardiological Hospitalizations (RECAH)
Deaths due to Respiratory (DDRE)

Table 5. Fuzzy Ruleset (Fuzzy-AND) for the determination of the Risk indices

ID	AirTemp	RH	SUN	MSMI Degree of Membership (DOM)	MPCI DOM	ID	AirTemp	RH	SUN	MSMI DOM	MPCI DOM
1	VL	VH	L	H(1)	L(0.1)	39	M	M	H	L(0.8)	M(0.2)
2	VL	VH	M	H(0.6)	L(0.2)	40	M	L	L	L(0.8)	M(0.2)
3	VL	VH	H	M(0.7)	L(0.3)	41	M	L	M	L(0.6)	M(0.5)
4	VL	H	L	H(0.8)	L(0.2)	42	M	L	H	L(0.4)	M(0.8)
5	VL	H	M	H(0.4)	L(0.3)	43	M	VL	L	L(0.7)	M(0.6)
6	VL	H	H	M(0.5)	L(0.4)	44	M	VL	M	L(0.5)	M(0.8)
7	VL	M	L	M(0.9)	L(0.4)	45	M	VL	H	L(0.3)	M(1)
8	VL	M	M	M(0.6)	L(0.5)	46	H	VH	L	M(0.3)	L(0.4)
9	VL	M	H	M(0.3)	L(0.6)	47	H	VH	M	M(0.2)	L(0.6)
10	VL	L	L	M(0.1)	L(0.6)	48	H	VH	H	M(0.1)	L(0.8)
11	VL	L	M	L(0.9)	L(0.7)	49	H	H	L	L(1)	L(0.5)
12	VL	L	H	L(0.5)	L(0.8)	50	H	H	M	L(0.9)	L(0.7)
13	VL	VL	H	L(0.4)	L(1)	51	H	H	H	L(0.8)	L(0.9)
14	VL	VL	M	L(0.7)	L(0.9)	52	H	M	L	L(0.9)	M(0.1)
15	VL	VL	L	L(1)	L(0.8)	53	H	M	M	L(0.8)	M(0.4)
16	L	VH	L	H(0.6)	L(0.5)	54	H	M	H	L(0.7)	M(0.7)
17	L	VH	M	H(0.3)	L(0.6)	55	H	L	L	L(0.8)	M(0.3)
18	L	VH	H	M(0.3)	L(0.7)	56	H	L	M	L(0.7)	H(0.3)
19	L	H	L	H(0.4)	L(0.6)	57	H	L	H	L(0.6)	H(0.4)
20	L	H	M	H(0.3)	L(0.7)	58	H	VL	L	L(0.7)	M(0.3)
21	L	H	H	M(0.3)	L(0.8)	59	H	VL	M	L(0.6)	H(0.3)
22	L	M	L	M(0.7)	L(0.7)	60	H	VL	H	L(0.5)	H(0.6)
23	L	M	M	M(0.4)	L(0.8)	61	VH	VH	L	L(1)	L(1)

(*continued*)

Table 5. (*continued*)

ID	AirTemp	RH	SUN	MSMI Degree of Membership (DOM)	MPCI DOM	ID	AirTemp	RH	SUN	MSMI DOM	MPCI DOM
24	L	M	H	M(0.1)	L(0.9)	62	VH	VH	M	L(0.9)	L(0.7)
25	L	L	L	L(0.9)	L(0.8)	63	VH	VH	H	L(0.8)	L(0.4)
26	L	L	M	L(0.7)	L(0.9)	64	VH	H	L	L(0.8)	L(0.5)
27	L	L	H	L(0.5)	L(1)	65	VH	H	M	L(0.7)	L(0.9)
28	L	VL	L	L(0.8)	M(0.1)	66	VH	H	H	L(0.6)	M(0.1)
29	L	VL	M	L(0.6)	M(0.2)	67	VH	M	L	L(0.6)	M(0.3)
30	L	VL	H	H(0.4)	M(0.3)	68	VH	M	M	L(0.5)	M(0.6)
31	M	VH	L	M(1)	L(0.3)	69	VH	M	H	L(0.4)	M(0.9)
32	M	VH	M	M(0.8)	L(0.5)	70	VH	L	L	L(0.4)	M(0.5)
33	M	VH	H	M(0.6)	L(0.7)	71	VH	L	M	L(0.3)	H(0.4)
34	M	H	L	M(0.8)	L(0.4)	72	VH	L	H	L(0.2)	H(0.8)
35	M	H	M	M(0.5)	L(0.6)	73	VH	VL	L	L(0.3)	M(0.7)
36	M	H	H	M(0.2)	L(0.8)	74	VH	VL	M	L(0.2)	H(0.6)
37	M	M	L	M(0.2)	L(0.8)	75	VH	VL	H	L(0.1)	H(1)
38	M	M	M	L(1)	L(1)						

Step4. For all seasons, we have examined the effect of the temporal delay of the Unified Risk index to the morbidity and mortality of the residents of Thessaloniki wider area (prefecture).

In stages 2 and 4 we have examined the 1–7 days temporal delay of the atmospheric conditions to the public health. In particular, the morbidity and mortality of circulatory and respiratory diseases are shifted from one day to seven days ahead (Table 6).

Table 6. Fuzzy Ruleset (Fuzzy-AND) for the determination of the Pollution Index due to Smog (PIDS)

ID	SO$_2$	PM$_{10}$	PM$_{2.5}$	CO	PIDS DOM	ID	SO$_2$	PM$_{10}$	PM$_{2.5}$	CO	PIDS DOM
1	L	L	L	L	L(1)	42	M	M	M	H	H(0.2)
2	L	L	L	M	L(0.9)	43	M	M	H	L	M(0.9)
3	L	L	L	H	L(0.6)	44	M	M	H	M	H(0.2)
4	L	L	M	L	L(0.9)	45	M	M	H	H	H(0.5)
5	L	L	M	M	M(0.2)	46	M	H	L	L	M(0.4)
6	L	L	M	H	M(0.4)	47	M	H	L	M	M(0.9)
7	L	L	H	L	L(0.6)	48	M	H	L	H	H(0.1)
8	L	L	H	M	M(0.4)	49	M	H	M	L	M(0.9)
9	L	L	H	H	M(0.7)	50	M	H	M	M	H(0.2)
10	L	M	L	L	L(0.9)	51	M	H	M	H	H(0.5)
11	L	M	L	M	M(0.2)	52	M	H	H	L	H(0.1)
12	L	M	M	M	M(0.8)	53	M	H	H	M	H(0.5)
13	L	M	M	H	M(0.9)	54	M	H	H	H·	H(0.8)

(*continued*)

Table 6. (*continued*)

ID	SO$_2$	PM$_{10}$	PM$_{2.5}$	CO	PIDS DOM	ID	SO$_2$	PM$_{10}$	PM$_{2.5}$	CO	PIDS DOM
14	L	M	H	L	M(0.4)	55	H	L	L	L	L(0.6)
15	L	M	H	M	M(0.9)	56	H	L	L	M	M(0.4)
16	L	M	L	H	M(0.4)	57	H	L	L	H	M(0.7)
17	L	'M	M	L	M(0.2)	58	H	L	M	L	M(0.4)
18	L	M	H	H	H(0.1)	59	H	L	M	M	M(0.9)
19	L	H	L	L	L(0.6)	60	H	L	M	H	H(0.1)
20	L	H	L	M	M(0.4)	61	H	L	H	L	M(0.7)
21	L	H	L	H	M(0.7)	62	H	L	H	M	H(0.1)
22	L	H	M	L	M(0.4)	63	H	L	H	H	H(0.4)
23	L	H	M	M	M(0.9)	64	H	M	L	L	M(0.4)
24	L	H	M	H	H(0.1)	65	H	M	L	M	M(0.9)
25	L	H	H	L	M(0.7)	66	H	M	L	H	H(0.1)
26	L	H	H	M	H(0.1)	67	H	M	M	L	M(0.9)
27	L	H	H	H	H(0.4)	68	H	M	M	M	H(0.2)
28	M	L	L	L	L(0.9)	69	H	M	M	H	H(0.5)
29	M	L	L	M	M(0.2)	70	H	M	H	L	H(0.1)
30	M	L	L	H	M(0.4)	71	H	M	H	M	H(0.5)
31	M	L	M	L	M(0.2)	72	H	M	H	H	H(0.8)
32	M	L	M	M	M(0.8)	73	H	H	L	L	M(0.7)
33	M	L	M	H	M(0.9)	74	H	H	L	M	H(0.1)
34	M	L	H	L	M(0.4)	75	H	H	L	H	H(0.4)
35	M	L	H	M	M(0.9)	76	H	H	M	L	H(0.1)
36	M	L	H	H	H(0.1)	77	H	H	M	M	H(0.5)
37	M	M	L	L	M(0.2)	78	H	H	M	H	H(0.8)
38	M	M	L	M	M(0.8)	79	H	H	H	L	H(0.4)
39	M	M	L	H	M(0.9)	80	H	H	H	M	H(0.8)
40	M	M	M	L	M(0.8)	81	H	H	H	H	H(1)
41	M	M	M	M	M(1)						

Fuzzy Chi Square Test was used for all stages per season to estimate the degree of dependency between the linguistics of the three indices and the linguistics of morbidity and mortality due to Cardiovascular and respiratory diseases, setting a confidence interval of 95%.

4   Results and Discussion

After extensive tests were conducted for all the seasons of the year, all possible cases of dependence between the L/FS of the three developed indices and the L/FS of the morbidity and mortality of cardiovascular and respiratory diseases were examined. Degrees of membership close to the value of 1, declare a higher level of correlation of the raw value to the fuzzy set (Linguistic). Many cases under consideration have attributed a high degree of membership (participation) to the fuzzy set corresponding to the Linguistic "strong dependence", stating that they represent to a satisfactory degree the reported fuzzy set.

This has been done in order to go one step further than the X square test. We not only wish to see if there is a relation (dependency or independency) but we mainly wish to see its level of strength. It is important to know if the X square test has offered a strong or a weak or a moderate dependency. This would help a lot the reasoning of the fuzzy inference system. For example, in this paper in many cases the degrees of membership to the fuzzy set "strong dependency" fall in the interval [0.7–1] which declares high dependency.

The most important dependence and strong correlation results are presented below:
Winter:

For the winter season, the Smog Pollution Index (PIDS) showed a high level of dependence (HLDEP) with a DOM equal to 0.90755 with cardiovascular deaths on the same day. It also proved to have HLDEP equal to 0.81575 with cardiovascular hospitalizations after a lag of 4 days.

Spring and autumn:

During the Spring-Autumn period the Meteorological risk index (METI) has proven to be highly related with a DOM 0.8637 with the respiratory deaths after a lag of 6 days. The pollution index (POLI) was highly related at a level equal to 0.9916 with the cardiovascular hospitalizations during the same day.

Summer:

For the summer, the MPCI (photochemical cloud meteorological index) was related to the respiratory hospitalizations with a DOM 0.69295 after a lag of 3 days (Table 7).

The PIDP (Pollution index related to photochemical cloud) was highly related to the cardiovascular deaths with a DOM equal to 0.83575 at the same day and it was also related to the respiratory hospitalizations with a DOM 0.7772 with a temporal delay of 4 days.

Winter:

During the winter the Unified risk index produced by the Min T-Norm (Min-URI) has shown an extremely HLDEP with a DOM 0.94755 with the cardiovascular deaths on the same day. Also the Min-URI appeared to have a high dependence with the cardiovascular and respiratory hospitalizations under a temporal lag of 4 to 6 days. The URI obtained by the Algebraic Product T-Norm (Alg_URI) had a high relation to the cardiovascular hospitalizations after a time lag of 4 days with a DOM equal to 0.70105.

Summer:

For the summer, the Alg_URI is highly related to the cardiovascular and respiratory morbidity and mortality not only of the same day but through a time lag of 7 days.

The URI produced by the Einstein T-Norm had a high dependence with a DOM 0.84595 with the cardiovascular hospitalizations after a time lag of 5 days.

The Hamacher URI was highly correlated to the cardiovascular and respiratory hospitalizations with a time delay of 3 days.

Table 7. Fuzzy Chi-Square Test between all Risk indices Linguistics and Linguistics of mortality and morbidity

Indices- Disease -Season	Statistic Test	P-Value	Linguistics of P-Value	Degree of Membership
MSMI - circulatory deaths (1 day lag) winter	8.0799	0.017599	MEDIUM	0.62995
MSMI - respiratory deaths (6 days lag) winter	4.8486	0.027668	MEDIUM	0.8666
PIDS - circulatory deaths winter	17.0991	0.001849	HIGH	0.90755
PIDS - circulatory deaths (4 days lag) winter	15.5508	0.003685	HIGH	0.81575
PIDS - respiratory hospitalization (4 days lag) winter	14.2136	0.006644	HIGH	0.6678
PIDS - respiratory deaths (7 days lag) winter	10.0136	0.006692	HIGH	0.6654
METI - circulatory deaths (3 days lag) spring& autumn	10.054	0.039594	LOW	0.4797
METI - respiratory deaths (6 days lag) spring& autumn	11.81	0.002726	HIGH	0.8637
POLI - circulatory hospitalization spring & autumn	22.3822	0.000168	HIGH	0.9916
POLI - circulatory hospitalization (1 day lag) spring & autumn	14.3268	0.006322	HIGH	0.6839
MPCI - respiratory hospitalization (3 days lag) summer	10.1855	0.006141	HIGH	0.69295
MPCI - respiratory hospitalization (6 days lag) summer	10.0138	0.006692	HIGH	0.6654
PIDP - circulatory deaths summer	8.6418	0.003285	HIGH	0.83575
PIDP - respiratory hospitalizations (3 days lag) summer	6.4048	0.011381	HIGH	0.43095
PIDP - respiratory hospitalizations (4 days lag) summer	8.0879	0.004456	HIGH	0.7772
PIDP - respiratory hospitalizations (5 days lag) summer	6.4691	0.010977	HIGH	0.45115
PIDP - circulatory deaths (5 days lag) summer	5.604	0.017919	MEDIUM	0.64595
PIDP - respiratory deaths (7 days lag) summer	6.0024	0.014287	MEDIUM	0.46435

*Meteorological Index METI
Pollution Index POLI

The Lukasiewicz URI had HLDEP with a DOM equal to 0.7799 with the cardiovascular hospitalizations with a temporal lag of 4 days and it was depended to the respiratory hospitalizations with a DOM 0.9085 at a time lag of 3 days (Table 8).

Table 8. Fuzzy Chi-Square Test of URI Linguistics and morbidity plus mortality for all seasons

Indices-Diseases-Season Winter	Statistic Test	P-Value	Linguistics of P-Value	Degree of membership
MIN - Deaths circulatory	18.3607	0.001049	HIGH	0.94755
MIN - Hospitalization circulatory (4 days lag)	15.8303	0.003255	HIGH	0.83725
MIN - Hospitalization respiratory (4 days lag)	14.2136	0.006644	HIGH	0.6678
MIN - Hospitalization circulatory (6 days lag)	13.6614	0.008458	HIGH	0.5771
MIN - Hospitalization respiratory (6 days lag)	13.0433	0.011066	HIGH	0.4467
MIN - Deaths respiratory (7 days lag)	9.7489	0.007639	HIGH	0.61805
Algebraic - Hospitalization circulatory (4 days lag)	10.2392	0.005979	HIGH	0.70105
Hamacher - Deaths respiratory (4 days lag)	5.7587	0.016407	MEDIUM	0.57035
Summer				
Min - Hospitalization respiratory (2 days lag)	7.1071	0.028622	MEDIUM	0.8189
Min - Deaths circulatory (6 days lag)	6.5369	0.038065	LOW	0.40327
Min - Hospitalization circulatory (6 days lag)	6.982	0.03047	MEDIUM	0.7265
Min - Hospitalization respiratory (6 days lag)	7.8491	0.019751	MEDIUM	0.73755
Min - Deaths respiratory (7 days lag)	7.8065	0.020176	MEDIUM	0.7588
Algebraic - Hospitalization circulatory	32.7397	<0.00001	HIGH	1
Algebraic - Hospitalization circulatory (1 day lag)	20.6015	<0.00001	HIGH	1
Algebraic - Hospitalization circulatory (4 days lag)	22.9634	<0.00001	HIGH	1
Algebraic - Hospitalization circulatory (5 days lag)	20.2869	<0.00001	HIGH	1
Algebraic - Hospitalization circulatory (6 days lag)	21.0242	<0.00001	HIGH	1

(continued)

Table 8. (*continued*)

Indices-Diseases-Season Winter	Statistic Test	P-Value	Linguistics of P-Value	Degree of membership
Algebraic - Hospitalization circulatory (7 days lag)	19.5166	<0.00001	HIGH	1
Algebraic - Hospitalization respiratory	25.7959	<0.00001	HIGH	1
Algebraic - Hospitalization respiratory (1 day lag)	22.7579	<0.00001	HIGH	1
Algebraic - Hospitalization respiratory (5 days lag)	20.2312	<0.00001	HIGH	1
Algebraic - Deaths circulatory	31.0503	<0.00001	HIGH	1
Algebraic - Deaths circulatory (1 day lag)	24.0131	<0.00001	HIGH	1
Algebraic - Deaths circulatory (2 days lag)	21.6494	<0.00001	HIGH	1
Algebraic - Deaths circulatory (4 days lag)	20.2473	<0.00001	HIGH	1
Algebraic - Deaths circulatory (5 days lag)	28.4996	<0.00001	HIGH	1
Algebraic - Deaths circulatory (7 days lag)	37.9859	<0.00001	HIGH	1
Algebraic - Deaths respiratory	18.944	0.000013	HIGH	0.99935
Algebraic - Deaths respiratory (1 day lag)	13.5957	0.000227	HIGH	0.98865
Algebraic - Deaths respiratory (2 days lag)	11.8405	0.000580	HIGH	0.971
Algebraic - Deaths respiratory (3 days lag)	14.3905	0.000149	HIGH	0.99255
Algebraic - Deaths respiratory (4 days lag)	12.9718	0.000316	HIGH	0.9842
Algebraic - Deaths respiratory (5 days lag)	16.4705	0.000049	HIGH	0.99755
Einstein - Deaths circulatory	4.3372	0.037288	LOW	0.3856
Einstein - Hospitalization circulatory (5 days lag)	8.7586	0.003081	HIGH	0.84595
Hamacher - Hospitalization circulatory (3 days lag)	7.1775	0.007382	HIGH	0.6309
Hamacher - Hospitalization circulatory (4 days lag)	5.4406	0.019674	MEDIUM	0.7337
Hamacher - Hospitalization respiratory (3 days lag)	23.2524	<0.00001	HIGH	1
Hamacher - Hospitalization respiratory (4 days lag)	10.0854	0.001495	HIGH	0.92525

(*continued*)

Table 8. (*continued*)

Indices-Diseases-Season Winter	Statistic Test	P-Value	Linguistics of P-Value	Degree of membership
Lukasiewicz - Hospitalization circulatory (5 days lag)	8.11	0.004402	HIGH	0.7799
Lukasiewicz - Hospitalization circulatory (6 days lag)	7.2356	0.007147	HIGH	0.64265
Lukasiewicz - Hospitalization respiratory (1 day lag)	5.8083	0.01595	MEDIUM	0.5475
Lukasiewicz - Hospitalization respiratory (3 days lag)	9.7127	0.00183	HIGH	0.9085
Spring & Autumn MIN - Hospitalization circulatory	32.1649	<0.00001	HIGH	1
MIN - Hospitalization circulatory (1 day lag)	18.0139	0.001226	HIGH	0.9387
MIN - Hospitalization circulatory (2 days lag)	14.4578	0.005969	HIGH	0.70155
Algebraic - Hospitalization circulatory	32.6394	<0.00001	HIGH	1
Algebraic - Hospitalization respiratory (5 days lag)	7.2267	0.026961	MEDIUM	0.90195
Algebraic - Deaths circulatory (5 days lag)	8.7982	0.003015	HIGH	0.84925
Einstein - Hospitalization circulatory	6.7215	0.034709	MEDIUM	0.51455
Einstein - Hospitalization respiratory (5 days lag)	6.6959	0.035156	MEDIUM	0.4922
Einstein - Deaths circulatory (5 days lag)	7.9346	0.00485	HIGH	0.7575
Hamacher - Deaths circulatory (2 days lag)	6.9904	0.008195	HIGH	0.59025
Hamacher - Deaths circulatory (3 days lag)	8.8459	0.002937	HIGH	0.85315
Hamacher - Deaths circulatory (4 days lag)	6.7173	0.009548	HIGH	0.5226
Lukasiewicz - Deaths circulatory (2 days lag)	6.8058	0.009086	HIGH	0.5457
Lukasiewicz - Deaths circulatory (3 days lag)	9.2065	0.002412	HIGH	0.8794
Lukasiewicz - Deaths circulatory (4 days lag)	7.3579	0.006677	HIGH	0.66615

Spring and autumn:

During the spring-autumn season the URI (of the Min and the Algebraic Product T-Norms) was extremely highly correlated with a DOM of 1 to the cardiovascular hospitalizations of the same day.

The Einstein URI was depended highly with a DOM 0.7575 with the cardiovascular deaths within a time lag of 5 days. Finally the Hamacher and Lukasiewicz URI had a very high correlation to the cardiovascular deaths at a time lag of 3 days.

Table 9 proves that for the summer all 5 URI obtained by 5 T-Norms are related to the cardiovascular hospitalizations at a level of 100%. In the spring-autumn season and also in the summer, 4 out of the 5 URI are highly related to both cardiovascular morbidity and mortality at a level of 80%. Also during the summer 4 out of the 5 URI are related to the respiratory morbidity whereas 2 out of the 5 URI are highly correlated to the respiratory mortality.

Table 9. Seasonal percentages of 5 T-Norms per season related to the morbidity and mortality due to circulatory and respiratory diseases

Season	Hospitalization circulatory	Hospitalization respiratory	Deaths circulatory	Deaths respiratory
Winter	40%	20%	20%	40%
Summer	100%	80%	80%	40%
Spring & Autumn	80%	40%	80%	0%

5 Conclusion

This paper proposes the use of an innovative holistic method of investigating meteorological parameters by season that maximize atmospheric pollution. The intelligent system developed was based on the combination of fuzzy methods with the hybrid non-parametric Fuzzy Chi-Square Test. The seasonal indices created were consolidated using five fuzzy conjunction T-norms in a combined Unifier risk index containing the hazard of all elements of the atmosphere.

The operation of this model calculated the degree of dependence between the produced fuzzy risk indices and the morbidity and mortality on the inhabitants of the prefecture of Thessaloniki. Thus, it indicated the correlation of the effects of the climatic conditions of the atmosphere on public health based on short time lags.

This modeling research effort has yielded high rates of correlations between the obtained indices and the cardiovascular-respiratory morbidity and mortality.

As future directions that could improve the proposed model, we suggest the potential use of more parameters directly related to morbidity and mortality of an urban center in order to create more combinations of fuzzy rules and sub-indicators. In this way it will be possible to derive an even stronger final adaptive unified index.

Finally, we suggest the future use of other machine learning methods (e.g. unsupervised or competitive learning) hybrid soft computing approaches (fuzzy-neural networks) and optimization algorithms.

References

1. Anderson, B.G., Bell, M.L.: Weather-related mortality: How heat, cold, and heat waves affect mortality in the United States. Epidemiology **20**(2), 205–213 (2009). doi:10.1097/EDE.0b013e318190ee08
2. Cakmak, S., Hebbern, C., Cakmak, J.D., Vanos, J.: The modifying effect of socioeconomic status on the relationship between traffic, air pollution and respiratory health in elementary schoolchildren. J. Environ. Manage. **177**, 1–8 (2016). doi:10.1016/j.jenvman.2016.03.051
3. Calvo, T., Mayor, G., Mesira, R.: Aggregation Operators: New Trends and Applications. Studies in Fuzziness and Soft Computing. Physica-Verlag, Heidelberg (2002)
4. Corder, G.W., Foreman, D.I.: Nonparametric Statistics: A Step-by-Step Approach. Wiley, Hoboken (2014). ISBN 978-1118840313
5. Cox, E.: Fuzzy Modeling and Genetic Algorithms for Data Mining and Exploration. Elsevier Science, USA (2005)
6. Greenwood, P.E., Nikulin, M.S.: A Guide to Chi-squared Testing. Wiley, New York (1996). ISBN 978-0-471-55779-1
7. Huang, C.E., Ruan, D., Kerre, E.: Fuzzy risks and updating a fuzzy risk with new Observations. J. Risk Anal. Int. J. (2007, in press)
8. Iliadis, L., Papaleonidas, A.: Computational Intelligence and Intelligent Agents (in Greek). Tziolas Edition, Thessaloniki (2016)
9. International Statistical Classification of Diseases and Related Health Problems 10th Revision. http://apps.who.int/classifications/icd10/browse/2016/en#
10. Kassomenos, P.A., Kelessis, A., Paschalidou, A.K., Petrakakis, M.B.: Identification of sources and processes affecting particulate pollution in Thessaloniki, Greece. Atmos. Environ. **45**(39), 7293–7300 (2011). doi:10.1016/j.atmosenv.2011.08.034
11. Kecman, V.: Learning and Soft Computing. MIT Press, London (2001)
12. Kyriakidis, I., Karatzas, K., Papadourakis, G., Ware, A., Kukkonen, J.: Investigation and forecasting of the common air quality index in Thessaloniki, Greece. In: Iliadis, L., Maglogiannis, I., Papadopoulos, H., Karatzas, K., Sioutas, S. (eds.) AIAI 2012. IAICT, vol. 382, pp. 390–400. Springer, Heidelberg (2012). doi:10.1007/978-3-642-33412-2_40
13. Leondes, C.: Fuzzy logic and Expert Systems Applications. Academic Press, San Diego (1998)
14. Nguyen, H., Walker, E.: A first Course in Fuzzy Logic. Chapman and Hall, Boca Raton (2000)
15. Papanastasiou, D.K., Melas, D., Kambezidis, H.D.: Air quality and thermal comfort levels under extreme hot weather. Atmos. Res. **152**, 4–13 (2015). doi:10.1016/j.atmosres.2014.06.002
16. Prasad, K., Gorai, A.K., Goyal, P.: Development of ANFIS models for air quality forecasting and input optimization for reducing the computational cost and time. Atmos. Environ. **128**, 246–262 (2016). doi:10.1016/j.atmosenv.2016.01.007
17. Sarigiannis, D.A., Karakitsios, S.P., Kermenidou, M.V.: Health impact and monetary cost of exposure to particulate matter emitted from biomass burning in large cities. Sci. Total Environ. **524–525**, 319–330 (2015). doi:10.1016/j.scitotenv.2015.02.108
18. The MathWorks. http://www.mathworks.com/
19. Tsangari, H., Paschalidou, A.K., Kassomenos, A.P., Vardoulakis, S., Heaviside, C., Georgiou, K.E., Yamasaki, E.N.: Extreme weather and air pollution effects on cardiovascular and respiratory hospital admissions in Cyprus. Sci. Total Environ. **542**, 247–253 (2016). doi:10.1016/j.scitotenv.2015.10.106

20. Vardoulakis, S., Dear, K., Hajat, S., Heaviside, C., Eggen, B., McMichael, A.J.: Comparative assessment of the effects of climate change on heat-and cold-related mortality in the United Kingdom and Australia. Environ. Health Perspect. **122**(12), 1285–1292 (2015). doi:10.1289/ehp.1307524

21. Vlachokostas, C., Achillas, C., Moussiopoulos, N., Kalogeropoulos, K., Sigalas, G., Kalognomou, E.A., Banias, G.: Health effects and social costs of particulate and photochemical urban air pollution: a case study for Thessaloniki, Greece. Air Qual. Atmos. Health **5**(3), 325–334 (2012). doi:10.1007/s11869-010-0096-1

22. Voukantsis, D., Karatzas, K., Kukkonen, J., Räsänen, T., Karppinen, A., Kolehmainen, M.: Intercomparison of air quality data using principal component analysis, and forecasting of PM10 and PM2.5 concentrations using artificial neural networks, in Thessaloniki and Helsinki. Sci. Total Environ. **409**(7), 1266–1276 (2011). doi:10.1016/j.scitotenv.2010.12.039

23. Xie, W., Li, G., Zhao, D., Xie, X., Wei, Z., Wang, W., Wang, M., Li, G., Liu, W., Sun, J., Jia, Z., Zhang, Q., Liu, J.: Relationship between fine particulate air pollution and ischaemic heart disease morbidity and mortality. Heart **101**(4), 257–263 (2015). doi:10.1136/heartjnl-2014-306165

24. Zoumakis, M., Kassomenos, P., Zoumakis, N., Papadakis, N., Vosniakos, F., Staliopoulou, M., Efstathiou, G., Kelessis, A., Papapreponis, P., Bournis, N., Petrakakis, M., Kozyraki, T. H.: Human discomfort due to environmental conditions in Urban Thessaloniki, Greece. Part V. Adaptation to the local climate (Preliminary results). J. Environ. Protect. Ecol. **12**(3), 921–929 (2011)

Understanding Decision Support in Large-Scale Disasters: Challenges in Humanitarian Logistics Distribution

Mohammad Tafiqur Rahman[1]([⊠]), Tina Comes[2], and Tim A. Majchrzak[1]

[1] University of Agder, Kristiansand, Norway
tafiqur.rahman@uia.no
[2] Delft University of Technology, Delft, Netherlands

Abstract. Disasters are characterized by conflicting, uncertain, or lacking data. Nevertheless, humanitarian responders need to make rapid decisions. This is particularly true for the immediate response to a sudden onset disaster. Since most humanitarian decision support systems (DSS) make important assumptions on data availability and quality that are often not fulfilled in practice, decision-makers are largely left to their experience. In this paper, we identify three major challenges for an operational DSS to support distribution planning: (i) deep uncertainty; (ii) reflecting field conditions and constraints; and (iii) rapid humanitarian logistics modeling. We review the relevant theories and provide an outline of the system requirements to develop a system for operational responders to achieve targeted service level on distribution of disaster relief through proper utilization of resources, time and scheduling.

Keywords: Decision support systems · Deep uncertainty · Disaster response · Rapid decision-making · Humanitarian logistics distribution

1 Introduction

For decades, responders have been searching for information and data to provide better aid. Today, there is no information shortage in disasters any more: increasing public participation, volunteers, responders and affected communities alike produce large amount of data [11,13]. However, such datasets are often incomplete because of infrastructure disruptions and access difficulties [14]. Even if data is available, it needs to be formatted and shared to support coordination [16] so that response can be organized in an efficient and effective way [49].

Humanitarian logistics becomes vital when natural disasters strike. Responders need to deal with uncertainties in the shortest possible time to make decisions that save human lives and alleviate suffering. However, humanitarian needs (demands) and available resources and goods (supplies) are dynamically evolving [57]. Particularly in the early phases of the response, little is known about what is needed at which place, and responders are left to make decisions without appropriate data or computational support [13].

© Springer International Publishing AG 2017
I.M. Dokas et al. (Eds.): ISCRAM-med 2017, LNBIP 301, pp. 106–121, 2017.
DOI: 10.1007/978-3-319-67633-3_9

This paper discusses challenges and approaches for developing a dynamic information system (IS) to support decision-making on distribution disaster aid goods. We base our work on the Dynamic Emergence Response Management Information Systems (DERMIS) principles, developed by Turoff et al. [51] to assist efficient and effective decision-making.

The remainder of this paper is divided into three parts. Firstly, we discuss different research challenges to identify gaps that require further research. Secondly, we present our research approach. Finally, we draw conclusions.

2   Research Challenges

Due to many socio-technical interdependencies, the onset of a disaster causes complex chains of interactions [10] that are unique [62] and cannot be predicted. These characteristics of a disaster make it difficult for responders to identify relevant information [10] to allocate adequate humanitarian logistics for rapid action.

According to Van Wassenhove [58], proper allocation of resources is one of the reasons for avoiding the creation of bottlenecks in relief operation and it requires management procedures to maintain information and resources for rapid and timely response [44]. Many authors conduct systematic reviews and survey on the aspect of relief goods distribution networks in response to disasters and summarizes some promising challenges in this area [2, 18, 25, 27–29, 36–38, 52, 59]. After having an in-depth study of these articles, we realize that distribution planning for disaster relief goods is not a standalone area of research. It is connected with few other areas and thus complex in nature. For example, responders need to know the demand, products' sourcing and procurement and transport arrangements before making plans on relief distribution.

Anaya-Arenas et al. [2], Holguín-Veras et al. [25], Kovács and Spens [27], and Van Wassenhove and Pedraza Martínez [59] focus on demand uncertainty. According to them, demand is always unpredictable at the time of disaster and it changes frequently. To make proper distribution of relief goods, responders need to assess fields' demand and make arrangement to collect those demandable products. They perform several complex activities to arrange those scarce resources. Most of the authors of the articles listed above emphasize proper humanitarian supply chain modeling because it covers a wide area of distribution planning starting from fund collection to effective and efficient distribution. Anaya-Arenas et al. [2] identify challenges in stock management and relocation, transportation, and coordination, collaboration, and communication among humanitarian supply chains. These challenges are also supported by Cozzolino [18], Kovács and Spens [28], Nolz et al. [36], and Van Wassenhove and Pedraza Martínez [59]. Additionally, type of the disaster as well as the geolocation of disastrous area and its political and social structures play important roles in relief distribution planning to that specific area [25, 27, 38].

According to Tzeng et al. [52], after a disaster occurrence, decisions on humanitarian support requires a rapid analysis of the situation even on the basis of limited and often incomplete information and Nolz et al. [37] suggests

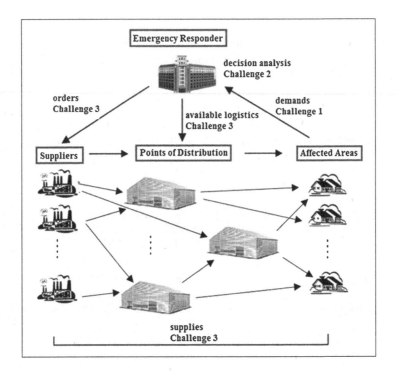

Fig. 1. Challenges in distribution of disaster aid goods

developing a decision support system to perform this complex analysis task and assist practitioners with alternative solutions. So, a computer based [45], well-structured [15], flexible [62], effective and efficient decision support system (DSS) is needed to acquire and analyze data from various sources and assist humanitarian logisticians to take decision on distribution of goods.

From the above discussion on challenges of disaster relief distribution, we conclude that for developing such DSS, researchers need to deal with the three most salient research challenges that are portrayed in Fig. 1. Challenge 1 deals with uncertain information on demand for humanitarian logistics; Challenge 2 analyzes that information to meet requested demand by taking field conditions and constraints into consideration; Challenge 3 discusses the humanitarian supply chain to make logistics available to the survivors.

2.1   Deep Uncertainty

The term *uncertainty* is present in (almost) every research field but differs in meaning and implication [53]. Walker et al. [53] and Pruyt and Kwakkel [42] define uncertainty as knowledge inadequacy or limitation on past, current and future events. Benini [7] discusses Smithson's (1989) typology of ignorance, where uncertainty is placed within a wider taxonomy of information ignorance under incomplete statements subclass (see Fig. 2) and it has its own subordinate types:

vague, ambiguous and probabilistic statements. Similarly, Walker et al. [53] identify that lack of knowledge or information is not the only reason for uncertainty; knowledge inexactness, unreliability and ignorance also brings uncertainty in the scene. Additionally, they claim that excessive information and latest or updated version of it can reveal the existence of uncertainty in certain situations. For example, inexact and unreliable information creates knowledge gap, whereas excess and new information makes current knowledge inconsistence and fragile. All of these information categories are widely available in disastrous events and thus, make situations uncertain. According to Luis et al. [33], uncertainty is prevalent in the distribution of humanitarian logistics and it exists throughout all phases of decision-making [14]. Pruyt and Kwakkel [42] divide uncertainty into seven levels: marginal uncertainty, shallow uncertainty, medium uncertainty, deep uncertainty, and recognized ignorance. The authors characterize deep uncertainty as the ability to generate or consider multiple possibilities with no tracing of internal relationship among themselves. Klibi et al. [26] and Comes et al. [14] define deep uncertainty as the absence of apposite information to asses and identify upcoming unusual events that need to be taken care of. Decision-making becomes more challenging when uncertainty is deep [14]. For example, when a disaster strikes, it creates a chaotic environment; the behavior of the complex socio-technical systems affect is unpredictable and the disruption of infrastructures causes breakdown in information flows [32]. Thus, it creates uncertainty in demand and supply of each relief item and in distribution networks due to possible damages to storage facilities and transportation links [50]. Pruyt and Kwakkel [42] identify three uncertainty prone situations:

1. if it is difficult to find appropriate mechanisms from many existing ones for real-world solution,
2. if the probabilities for acceptable real-world outcomes are uncertain and
3. if acceptance fluctuates.

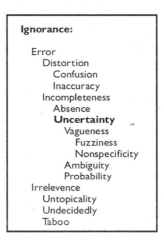

Fig. 2. Smithson's typology of ignorance, retrieved from [7]

According to Liberatore et al. [32], survivors and disastrous situations are not the only reasons for uncertainty. They claim that reporters, NGOs, and other emergency agencies are also responsible for making situations unstable. The information they provide is sometimes unstructured and cannot be utilized directly for identifying hidden relationships among datasets to assist decision-making in highly uncertain environment [31,32,54]. So, for making better decisions, uncertainty analysis becomes prominent, especially in disaster management [1].

Many researchers have been working on tackling uncertainties through scenario and model development. However, according to Hamarat et al. [21], no model or scenario should be considered as the *best* because the characteristics of uncertainties vary depending on the situations. Thus, decision makers always need some adjustments in decision-making variables in any system. Luis et al. [33] identify a trend on modelling uncertainty based on physical damages caused by a disaster and the immediate post-disaster functionalities. They point out uncertainty in demand and supply of relief goods. If the disaster is transboundary (e.g. across countries), it becomes more challenging to regain control [3].

Hamarat et al. [21] propose an iterative computational model-based approach for taking decision under deep uncertainty and demonstrate it with policy-making for energy transition. Their approach iteratively explores thousands of plausible scenarios to get a satisfying dynamic adaptive policy, where the concentration on time (rapidness) is missing. Comes et al. [14] emphasize time and availability of experts in their integrated framework to produce plausible scenarios. Thus, the authors advice future researchers to focus on the relevance of information to the decision type in the time available. This is relevant to the disastrous events, where getting correct information in time is crucial. Thus, Pruyt and Kwakkel [42] generalize the approach to deal with deep uncertainty. According to them, alternative outcomes should be generated by using multiple possible features or alternative model structures. Figure 3 portraits different parameters that Liberatore et al. [32] suggest to consider while dealing with uncertainty in humanitarian logistics. As soon as data acquisition on demand (in deep uncertain situations) is completed (up to the considerable level), responders conduct decision analysis (Challenge 2) to identify the actual need.

2.2   Decision Analysis for Operational Response

Decision support technology is a generalized term that requires different systems, hardware and communication tools to work together with a purpose of solving real-time complex problems [49]. It requires integrated efforts from both humans and technical tools for proper decision-making in humanitarian support [4,45,47]. More specifically, decision support system is an integration of hardware and software, developed to assist humans on decision making [54]. [54] have identified four basic components of a decision sup-port system: (1) data repository, (2) data analysis capability, (3) models for solution, evaluation (alternative solutions) and recommending actions to take, and (4) technology for interactive usage of the data and model and to display them when necessary.

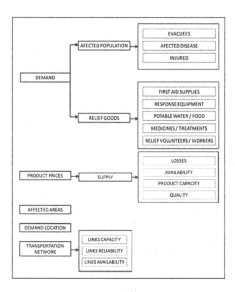

Fig. 3. Uncertainty parameters in humanitarian logistics [32]

Thompson et al. [49] note three sources of data inflow during a disaster: public, news media, and reports. According to the authors, public data is unreliable (to some degree), news media only covers a specific location and initial reports from public safety organizations alter by current situations, available resources and freelance responses. So, an appropriate decision support system should be able to integrate data (structured and unstructured) from multiple sources, analyze them and make them ready to use by the normative models for producing solutions and thus, can help practitioners to mitigate problems in relief distribution.

Shim et al. [45] propose a simplified decision-making process model that includes seven phases (see Fig. 4): problem recognition, problem definition, alternative generation, model development, alternative analysis, choice and implementation. So, decision analysis is necessary to decompose complicated problems into smaller and manageable components for decision aid with potential choices [46] that reflect the desire of the decision-maker [41]. However, decision analysis is all about explaining the decision situation and suggesting decision-making process [40], where computer system is integrated to facilitate these activities [46]. Balcik et al. [5] argue that it is necessary to consider the type of disaster, its impact on the affected society and location while designing decision support systems because it may cause relief activities to vary. Additionally, humanitarian operations are being affected by the realistic field conditions and constraints: lack of time, pressure to make decisions, lack of resources and often computational power, long working hours for volunteers, tensed atmosphere etc.

Developing an efficient and effective DSS for an operational responder is not an easy task. While working on a disastrous event, responders face many critical challenges. According to Thompson et al. [49], these challenges for humanitarian

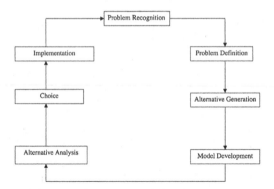

Fig. 4. Decision-making process of decision support systems [45]

response come from lack of coordination among responding actors, poor information flows between headquarter and field, and processing and validating relevant information for rapid decision-making. Thus, information becomes a vital resource for responders because it helps them to communicate with suppliers, coordinate with other agencies and allocate resources. Operational responders always suffer from information inaccessibly, lacking and overhead. According to Zang et al. [63], diversity of information and its distributed ownership are the major challenges in disaster relief operations. For example, when a disaster strikes, information can come from various sources: survivors, media and reports, and NGOs and other emergency agencies [32]. Thus, operational responders need to closely observe the situation, accumulate adequate and relevant information, properly analyze them and make an effective and efficient distribution plan [63].

According to Ter Mors et al. [48], operational responders should be able to prioritize tasks and identify relationships among them. This identification of relationship or dependency among tasks will help responders to determine who to communicate and coordinate to complete the tasks. Furthermore, operational responders should know when a task should be performed and under which circumstances [48]. While dealing with such complex tasks of resource allocation and scheduling, operational responders also need to concentrate on disrupted infrastructures, so that alternate distribution network can be planned. For achieving better control over the emergency response operations, Van de Walle and Turoff [61] emphasize the sensible development of flexible and dynamic operational decision support system. This system would facilitate operational responders to easily extract valid, relevant, meaningful and timely information. Nevertheless, this efficient extraction largely depends on quality and availability of inflow data and the effectiveness of data collection and analysis techniques [49]. After finalizing relief materials, their destination and time, emergency responders take initiative to find out how to meet the demand (Challenge 3).

2.3   Rapid Humanitarian Logistics Modeling

The concept of humanitarian supply chain (HSC) is different from the concept of commercial supply chain (CSC) and they differ from many aspects, such as assessing demand, sourcing and procurement, and delivery or supply infrastructure [30]. CSC has some structured ways to deal with demand and supply but HSC must deal with uncertainty in every phase of its functionality. For example, information inaccessibility, incompleteness, lacking and overload, unpredictable demand, disrupted infrastructure, damaged roads, delivery in short time and many other unknowns and uncertainties [30]. The general objective of humanitarian aids supply chain is to provide food, water, medicine, shelter and supplies immediately to people in need [6]. In humanitarian relief operations, needs and supplies are badly unstable because it is hard to predict demandable products and find potential suppliers [57]. Such operations require sufficient skills and knowledge on mobilizing disaster affected people and resources [58]. Thus, these operations demand proper communication and coordination among involved organizations and all the people within these organizations [3] to lessen the uncertainty in disaster relief distribution. Holguín-Veras et al. [25] claim that humanitarian logistics covers a broader working area in operational environment, where it always deals with urgent needs, life-or-death decisions and scarce of resources. This chaotic setting of operational environment becomes acute when transportation and communication networks collapse and these situations are quite normal when a disaster strikes. Thus, humanitarian supply chain should be managed effectively and efficiently to perform a successful relief operation [29]. Effectiveness ensures the targeted service level has been met and efficiency ensures that the response has been made by using minimum resources, cost and time. For achieving such efficient and effective humanitarian supply chain, Holguín-Veras et al. [25] suggest focusing on seven major components: (1) the objectives being pursued, (2) the origin of the commodity flows to be transported, (3) knowledge of demand, (4) the decision-making structure, (5) periodicity and volume of supply, and (6) the state of the social networks and (7) supporting systems.

Nevertheless, rapid response to the need of urgent relief goods is vital for saving victims' lives and humanitarian logistics plays an important role in such situations. According to Holguín-Veras et al. [24], it is common for humanitarian logistics to perform its activities when needs are at peak level, resources are insufficient and supporting systems are impacted. While responding to disastrous events, humanitarian logistics takes initiatives to search and rescue victims, and transport and distribute resources (equipment, personnel and goods) to assess the situation and meet the demand [25]. In this research, we are concentrating on rapid distribution of humanitarian relief goods that mostly includes faster demand assessment, quick allocation of resources and distribute them at an efficient and effective way. Thus, we need to model a humanitarian supply chain that would rapidly perform this activities for getting better control over emergency situations. In addition to placing demand and receiving supplies, the model should provide appropriate and updated stock information to operational responders. For better performance in distribution activities, the model should also be able to suggest alternative transportation networks (e.g. from supplier to warehouse, warehouse to warehouse, warehouse to the affected area). We consider that further studies are necessary to develop the concept of alternative distribution network along with the plan for time efficient and cost effective distribution of humanitarian logistics. For developing such a distribution model, we need to consider the following requirements [12, 20, 23]:

1. additional uncertainty (infrastructure disruption, security issues, political environment, facility location identification, unpredictable demands),
2. complex communication and coordination (information inaccessibility, shortage of logistics experts, multiple and unknown suppliers, local influential persons, government and civilians),
3. harder-to-achieve efficient and timely delivery (improper planning, lack of technology, manual processing, prioritization of goods, costly and lengthy distribution network selection), and
4. limited resources (funds, volunteers, vehicles, availability of relief items).

In addition to make decisions on products and suppliers, managers of humanitarian organization need to focus on procurement planning to obtain resources (donations, volunteers etc.), transport (from donor to beneficiary), warehousing (central and local), tracking and tracing for delivery confirmation [34].

Our research agenda is to develop an IS artifact (DSS) for supporting rapid humanitarian logistics decision-making on the basis of excessive, incomplete, conflicting and uncertain information (*constraints*). The aim of our DSS is to provide effective (reaching target service level), efficient (minimizing use of resources, cost and time) and equitable plans (facility location, flow mapping, routing and scheduling) for humanitarian logistics distribution. To support decision-making under uncertainties, the system will map current information flows against a series of bench-marks and thus determine the level of uncertainty along with various scenarios and adaptive paths. Thus, we will also focus on possible supply planning (especially the downstream part) for the prioritized distribution of disaster relief items as decision-makers not only have to decide which aids to deliver but also where and at what point in time. For developing such computerized IS artifact (DSS), we follow the related information categorization technique of Comes and Van de Walle [13]: (1) information on disaster that focuses on what (needs)-, where (infrastructures and accessibility)- and how (resources)- to support and (2) humanitarian logistics supply chain that deals with the distribution of relief materials and with all actors who are involved in this process.

3   Towards Rapid Humanitarian Distribution Planning – Our Approach

Designing an IS artifact will be designing IT product (DSS in our case) using information from appropriate sources and technologies by following explicit working procedures [55]. Depending on user requirements, system specifications and effective development, various design theories are available for different information systems types, such as management information systems (MIS), decision support systems (DSS), executive information systems (EIS), transaction processing systems (TPS), etc. [56]. According to Markus et al. [35], these different system types are considered as contributions to the IS field because their design theories and principles are developed basing on individual contexts.

In disaster management, it is necessary to perform complex tasks to deal with high uncertainty and multiple stakeholders and suppliers for providing rapid response to affected areas [39]. Human decision makers face difficulties to conduct these complex tasks for producing precise, efficient and unbiased outcomes [17]. They require computer aided DSS to process (such as order, store, retrieve, analyze) available information and share it among participating responders. According to Van de Walle and Turoff [61], this collaborative and integrative organizational work plays a key role to make a DSS successful. In addition to this sharing attitude, each DSS must follow a set of principles to assist decision-making on humanitarian logistics in feasible and effective ways [35,39,56].

Our research on rapid humanitarian distribution planning is in an emerging stage. Although we need more study to finalize our approach, here we present our generalized research steps to address the challenges depicted in the previous section. The development of our decision support system requires both empirical and design-oriented research. Thus, our work will follow a twofold approach:

1. Analyze how IS support decision-making on relief distribution planning in sudden onset disasters, and
2. Design, develop, implement, test and evaluate the decision support system.

The process model portrayed in Fig. 5 visualizes our research approaches. Initially, we develop our knowledge by reviewing existing DSS. We shall scrutinize those to understand how fast they support decision-making in specific problem domains (prioritizing logistic) by utilizing what type of information (structured, unstructured and/or sparse), technology and mechanism. This knowledge will help us identify the behavior of our targeted DSS. Then, we shall proceed to the design and development phase of our system. Finally, our system will be validated and empirically tested by some measurement instruments.

To develop the system, our research will deal with structured *and* unstructured data. The former includes data from topography, population registers, infrastructure, and pre-disaster functionality. The latter comes from sources ranging from sensor data to social media. There will be two data collection phases:

1. Design phase (or *first* evaluation cycle): data related to the targeted disasters from multi-disciplinary contexts will be acquired from social media, preparation packages, interview transcripts, observational notes-pictures-videos, surveys and online/offline documents.

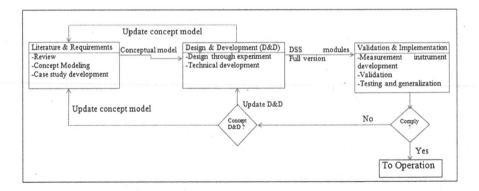

Fig. 5. Process model for our research approach

2. Evaluation phase (or *second* evaluation cycle): data will be collected during externally aligned project exercises and potentially complementary humanitarian training, also offering an opportunity to collect quantitative data from tables, decision points and duration, preferences, logistics flows etc.

For data analysis, we combine an quantitative and an qualitative approach. This mixed-method procedure will facilitate a complete and synergistic utilization of data to create rich insights on the targeted phenomenon [19]. For example, qualitative interview transcripts can be exploited to identify numeric data for statistical analysis, whereas quantitative survey data can be converted to descriptive notes for thematic or pattern analysis [9]. This indicates that numeric trends of qualitative data can be merged with the specific details received from quantitative data to identify new behavior for the system [22]. Venkatesh et al. [60] identify dominant and priority trends in this mixed-method of data analysis: quantitative dominated qualitative analysis or vice versa, where researchers collect data rigorously and analysis for domination. However, researchers are encouraged to put equal weight and emphasis on both data types.

According to Bhimani and Song [8] and Salvadó et al. [43], there is always a gap between practice on humanitarian operations and academic research because academic research does not have complete understanding on the needs in practice [8]. By considering the findings by [8,43], we plan to work collaboratively with the practitioners (through the aligned project) and involve them into the research from the very beginning. This will help us to understand their needs in detail and work accordingly. In this way, the evaluation process of our research can be started at early stage by the practitioners, which will facilitate them to check whether it is working effectively (reaching the targeted service level), efficiently (minimizing use of resources, cost and time) and equitably (dealing with facility location, prioritizing, flow mapping, alternate supply network, routing and scheduling) for the expected humanitarian logistics distribution operations.

4   Discussion and Conclusion

After pinning down the areas of concern for our research, we are focusing on developing a DSS that will rapidly analyze situations and assist stakeholders or practitioners on decision-making despite of lacking, conflicting and uncertain information. Moreover, we shall propose possible downward supply planning for effective, efficient and equitable distribution of humanitarian logistics to the affected areas. Since network decisions are strategic and often irreversible, temporal factors will also play an important role in our system: decision makers are required to choose not only *how* to set up a network, but also *when* there is sufficient information to make such a decision. Similarly, they do not only have to decide *which* aids to deliver but also *where* and at *what* point in time.

The success of our DSS depends on how it tackles the identified challenges: deep uncertainty, decision analysis for operational response and rapid humanitarian supply chain modeling. The demand for humanitarian logistics always fluctuates, which makes the entire system unstable. Humanitarian supply chains mostly deal with real-life constraints, such as suppliers, goods, delivery vehicles, roadways, warehouses, security, and volunteers. These physical constraints are crucial for distribution planning and impossible to upgrade at the time of disaster. However, we can switch from one provider to another for achieving targeted service levels on distribution of disaster relief through proper utilization of resources, time and scheduling. Emergency responders can switch after analyzing available information. So, we mainly target to upgrade the decision analysis phase in our research. Developing a system with the capability of rapid analysis of the demand and suggestion of alternatives (arrangements) for its delivery will accelerate distribution and, thus, respond to the emergency call quickly.

This decision support artifact will have specific user requirements, system specifications, and effective development procedures suitable for the contexts or scenarios that we will be concentrating on. In such way, it leads us to generate scenario-specific principles and build theories for designing and developing a DSS for humanitarian logistics distribution. Thus, it will create knowledge on certain problem settings and add values to IS research on decision support in humanitarian logistics. Nevertheless, this system will require adequate and quality information to produce proper decision. Additionally, the literature we studied for this research is mainly based on the Norwegian repository of Google Scholar (scholar.google.no), which limits our knowledge development on the context.

Our aimed decision-making support system will integrate and analyze information (structured and unstructured) from different heterogeneous sources to assist stakeholders for responding rapidly to disastrous situations with necessary humanitarian logistics, despite of time lacking for collecting additional information. Consequently, it is our research quest to find the best quality of information for generating this rapid decision support on humanitarian logistics distribution. Thus, the quality of the data stored in databases of various humanitarian organizations could be a challenge to deal with. Interviewing humanitarian assistance practitioners will be another challenging task to conduct. The research challenges and approaches we have discussed are a preliminary observation to design and develop our intended DSS. In near future, we will continue our literature work in other repositories and analyze the system requirements in detail to design the prototype, develop the system and test it in practice and thus, minimize the gap between academic research and practitioners' work.

References

1. Altay, N., Labonte, M.: Challenges in humanitarian information management and exchange: evidence from Haiti. Disasters **38**(s1), 50–72 (2014)
2. Anaya-Arenas, A.M., Renaud, J., Ruiz, A.: Relief distribution networks: a systematic review. Ann. Oper. Res. **223**(1), 53–79 (2014)
3. Ansell, C., Boin, A., Keller, A.: Managing transboundary crises: identifying the building blocks of an effective response system. JCCM **18**(4), 195–207 (2010)
4. Arnott, D., Pervan, G.: Eight key issues for the decision support systems discipline. Decis. Supp. Syst. **44**(3), 657–672 (2008)
5. Balcik, B., Beamon, B.M., Smilowitz, K.: Last mile distribution in humanitarian relief. J. Intell. Transp. Syst. **12**(2), 51–63 (2008)
6. Beamon, B.M., Balcik, B.: Performance measurement in humanitarian relief chains. Int. J. Publ. Sector Manage. **21**(1), 4–25 (2008)
7. Benini, A.A.: Uncertainty and information flows in humanitarian agencies. Disasters **21**(4), 335–353 (1997)
8. Bhimani, S., Song, J.S.: Gaps between research and practice in humanitarian logistics. J. Appl. Bus. Econ. **18**(1), 11–24 (2016)
9. Caracelli, V.J., Greene, J.C.: Data analysis strategies for mixed-method evaluation deigns. Educ. Eval. Policy Anal. **15**(2), 195–207 (1993)
10. Chan, J., Comes, T.: Innovative research design-a journey into the information typhoon. Procedia Eng. **78**, 52–58 (2014)
11. Carver, L., Turoff, M.: Human-computer interaction: the human and computer as a team in emergency management information systems. Commun. ACM **50**(3), 33–38 (2007)
12. Caunhye, A.M., Nie, X., Pokharel, S.: Optimization models in emergency logistics: a literature review. Socio Econ. Plan. Sci. **46**(1), 4–13 (2012)
13. Comes, T., Van De Walle, B.: Information systems for humanitarian logistics: concepts and design principles. In: Haavisto, I., Spens, K., Kovcs, G. (eds.) Supply Chain Management for Humanitarians, pp. 257–284. Kogan Page, London (2016)
14. Comes, T., Hiete, M., Schultmann, F.: A decision support system for multi-criteria decision problems under severe uncertainty. J. Multi Criteria Decis. Anal. **20**, 28–49 (2013)
15. Comes, T., Hiete, M., Wijngaards, N., Schultmann, F.: Decision maps: a framework for multi-criteria decision support under severe uncertainty. Decis. Supp. Syst. **52**(1), 108–118 (2011)
16. Comes, T., Wijngaards, N., Van de Walle, B.: Exploring the future: runtime scenario selection for complex and time-bound decisions. Technol. Forecast. Soc. Change **97**, 29–46 (2015)
17. Comfort, L.K., Sungu, Y., Johnson, D., Dunn, M.: Complex systems in crisis: anticipation and resilience in dynamic environments. JCCM **9**(3), 144–158 (2001)
18. Cozzolino, A.: Humanitarian logistics and supply chain management. In: Cozzolino, A. (ed.) Humanitarian Logistics, pp. 5–16. Springer, Heidelberg (2012)

19. Creswell, J.W., Fetters, M.D., Ivankova, N.V.: Designing a mixed methods study in primary care. Ann. Family Med. **2**(1), 7–12 (2004)
20. Fritz Institute, Logistics and the Effective Delivery of Humanitarian Relief. http://www.fritzinstitute.org/PDFs/Programs/tsunamiLogistics0605.pdf. Accessed 10 Aug 2016
21. Hamarat, C., Kwakkel, J.H., Pruyt, E.: Adaptive robust design under deep uncertainty. Technol. Forecast. Soc. Change **80**, 408–418 (2013)
22. Hanson, W.E., Creswell, J.W., Clark, V.L.P., Petska, K.S., Creswell, J.D.: Mixed methods research designs in counseling psychology. J. Counsel. Psychol. **52**(2), 224 (2005)
23. Hellingrath, B., Widera, A.: Survey on major challenges in humanitarian logistics. In: Proceedings of the 8th International ISCRAM Conference, pp. 1–5 (2011)
24. Holguín-Veras, J., Pérez, N., Jaller, M., Van Wassenhove, L.N., Aros-Vera, F.: On the appropriate objective function for post-disaster humanitarian logistics models. J. Oper. Manage. **31**(5), 262–280 (2013)
25. Holguín-Veras, J., Jaller, M., Van Wassenhove, L.N., Pérez, N., Wachtendorf, T.: On the unique features of post-disaster humanitarian logistics. J. Oper. Manage. **30**(7), 494–506 (2012)
26. Klibi, W., Martel, A., Guitouni, A.: The design of robust value-creating supply chain networks: a criti-cal review. Eur. J. Oper. Res. **203**(2), 283–293 (2010)
27. Kovács, G., Spens, K.: Identifying challenges in humanitarian logistics. IJPDLM **39**(6), 506–528 (2009)
28. Kovács, G., Spens, K.M.: Humanitarian logistics in disaster relief operations. IJPDLM **37**(2), 99–114 (2007)
29. Kunz, N., Reiner, G.: A meta-analysis of humanitarian logistics research. J. Humanit. Logist. Supply Chain Manage. **2**(2), 116–147 (2012)
30. Lee, Y.M., Ghosh, S., Ettl, M.: Simulating distribution of emergency relief supplies for disaster response operations. In: Winter Simulation Conference, pp. 2797–2808 (2009)
31. Lempert, R.J., Groves, D.G., Popper, S.W., Bankes, S.C.: A general, analytic method for generating robust strategies and narrative scenarios. Manage. Sci. **52**(4), 514–528 (2006)
32. Liberatore, F., Pizarro, C., de Blas, C.S., Ortuño, M.T., Vitoriano B.: Uncertainty in humanitarian logistics for disaster management: a review. In: Vitoriano, B., Montero, J., Ruan, D. (eds) Decision Aid Models for Disaster Management and Emergencies. Atlantis Computational Intelligence Systems, vol. 7. Atlantis Press, Paris (2013)
33. Luis, E., Dolinskaya, I.S., Smilowitz, K.R.: Disaster relief routing: integrating research and practice. Socio Econ. Plan. Sci. **46**(1), 88–97 (2012)
34. Maon, F., Lindgreen, A., Vanhamme, J.: Supply Chains in Disaster Relief Operations: Cross-sector Socially Oriented Collaborations. Hull University Business School (2009)
35. Markus, M.L., Majchrzak, A., Gasser, L.: A design theory for systems that support emergent knowledge processes. MIS Q. **26**, 179–212 (2002)
36. Nolz, P.C., Semet, F., Doerner, K.F.: Risk approaches for delivering disaster relief supplies. OR Spectrum **33**(3), 543–569 (2011)
37. Nolz, P.C., Doerner, K.F., Hartl, R.F.: Water distribution in disaster relief. IJPDLM **40**(8/9), 693–708 (2010)
38. Oloruntoba, R.: A wave of destruction and the waves of relief: issues, challenges and strategies. Disaster Prevention Manage. **14**(4), 506–521 (2005)

39. Ortuño, M.T., Cristóbal, P., Ferrer, J.M., Martín-Campo, F.J., Muoz, S., Tirado, G., Vitoriano, B.: Decision aid models and systems for humanitarian logistics: a survey. In: Vitoriano, B., Montero, J., Ruan, D. (eds.) Decision Aid Models for Disaster Management and Emergencies. Atlantis Computational Intelligence Systems, vol. 7. Atlantis Press, Paris (2013)
40. Pawlak, Z.: Rough set approach to knowledge-based decision support. Eur. J. Oper. Res. **99**(1), 48–57 (1997)
41. Pitz, G.F.: Human engineering of decision aids. Adv. Psychol. **14**, 205–221 (1983)
42. Pruyt, E., Kwakkel, J.H.: Radicalization under deep uncertainty: a multi-model exploration of activism, extremism, and terrorism. Syst. Dyn. Rev. **30**, 1–28 (2014)
43. Salvadó, L.L., Lauras, M., Comes, T., Van de Walle, B.: Towards more relevant research on Humanitarian disaster management coordination. In: Proceedings of 12th ISCRAM (2015)
44. Shaluf, I.M., Ahmadun, F.R., Mustapha, S.A.: Technological disaster's criteria and models. Disaster Prevention Manage. **12**(4), 305–311 (2003)
45. Shim, J.P., Warkentin, M., Courtney, J.F., Power, D.J., Sharda, R., Carlsson, C.: Past, present, and future of decision support technology. Decis. Support Syst. **33**(2), 111–126 (2002)
46. Stabell, C.B.: Decision support systems: alternative perspectives and schools. Decis. Support Syst. **3**(3), 243–251 (1987)
47. Soden, R., Budhathoki, N., Palen, L.: Resilience-building and the crisis informatics agenda: lessons learned from open cities Kathmandu. In: Proceedings of 11th ISCRAM, University Park, Pennsylvania, USA, pp. 1–10 (2014)
48. Ter Mors, A., Valk, J., Witteveen, C.: An event-based task framework for disaster planning and decision support. In: 2nd ISCRAM, Brussels, Belgium (2005)
49. Thompson, S., Altay, N., Green, W.G., Lapetina, J.: Improving disaster response efforts with decision support systems. Int. J. Emerg. Manage. **3**(4), 250–263 (2006)
50. Tofighi, S., Torabi, S.A., Mansouri, S.A.: Humanitarian logistics network design under mixed uncertainty. Eur. J. Oper. Res. **250**(1), 239–250 (2016)
51. Turoff, M., Chumer, M., Van de Walle, B., Yao, X.: The design of a dynamic emergency response management information system (DERMIS). JITTA J. Inf. Technol. Theory Appl. **5**(4), 1 (2004)
52. Tzeng, G.H., Cheng, H.J., Huang, T.D.: Multi-objective optimal planning for designing relief delivery systems. Transp. Res. Part E Logist. Transp. Rev. **43**(6), 673–686 (2007)
53. Walker, W.E., Lempert, R.J., Kwakkel, J.H.: Deep uncertainty. In: Gass, S.I., Fu, M.C. (eds.) Encyclopedia of Operations Research and Management Science, pp. 395–402. Springer, New York (2013)
54. Wallace, W.A., De Balogh, F.: Decision support systems for disaster management. Publ. Admin. Rev. **45**, 134–146 (1985)
55. Walls, J.G., Widmeyer, G.R., El Sawy, O.A.: Assessing information system design theory in perspective: how useful was our 1992 initial rendition? JITTA J. Inf. Technol. Theory Appl. **6**(2), 43 (2004)
56. Walls, J.G., Widmeyer, G.R., El Sawy, O.A.: Building an information system design theory for vigilant EIS. Inf. Syst. Res. **3**(1), 36–59 (1992)
57. Widera, A., Dietrich, H.A., Hellingrath, B., Becker, J.: Understanding humanitarian supply chains developing an integrated process analysis toolkit. In: Proceedings of the 10th International ISCRAM Conference, Germany (2013)
58. Van Wassenhove, L.N.: Humanitarian aid logistics: supply chain management in high gear. J. Oper. Res. Soc. **57**(5), 475–489 (2006)

59. Van Wassenhove, L.N., Pedraza Martinez, A.J.: Using OR to adapt supply chain management best practices to humanitarian logistics. Int. Trans. Oper. Res. **19**(1–2), 307–322 (2012)
60. Venkatesh, V., Brown, S.A., Bala, H.: Bridging the qualitative-quantitative divide: guidelines for conducting mixed methods research in information systems. MIS Q. **37**(1), 21–54 (2013)
61. Van de Walle, B., Turoff, M.: Decision support for emergency situations. IseB **6**(3), 295–316 (2008)
62. Yates, D., Paquette, S.: Emergency knowledge management and social media technologies: a case study of the 2010 Haitian earthquake. Int. J. Inf. Manage. **31**(1), 6–13 (2011)
63. Zhang, D., Zhou, L., Nunamaker, J.F.: A knowledge management framework for the support of decision making in humanitarian assistance/disaster relief. Knowl. Inf. Syst. **4**(3), 370–385 (2002)

Decision Making in Disaster Management: From Crisis Modeling to Effective Support

Wissem Eljaoued[(✉)] and Narjès Bellamine Ben Saoud[(✉)]

RIADI Laboratory, National School of Computer Sciences,
University of Manouba, Manouba, Tunisia
wissem.eljaoued@ensi-uma.tn, Narjes.Bellamine@ensi.rnu.tn

Abstract. Crisis management is a challenging domain to model because of its multiple stakeholders, its complex resource management, its rich communication and control mechanisms, etc. It consists in sets of inter-dependent activities occurring in an evolving environment under stressful conditions and high pressures. It becomes difficult to support given that any decision includes multiple actors, multiple emergency services, etc. [1]. Therefore, we believe that a systemic modeling of such complex socio-technical system (including the decisions makers, stakeholders, coordination processes, etc.) would improve the potential to support effectively the crisis management. We focus in this paper on the effective use of modeling to manage crisis. In this paper, we present firstly most relevant existing metamodels of crisis management. Secondly, we describe the criteria we defined to classify these metamodels. Thirdly, we present the results of this classification and finally, we conclude by suggesting first requirements needed for an effective metamodel useful to support decision processes in crisis management.

Keywords: Decision making · Crisis management · Systemic modeling/metamodeling

1 Introduction

Crisis management is a challenging domain to model [1] because of a large number of activities, dynamic decision support requirements, changing environmental situations and multi-data sources [2]. To avoid human and material losses in disaster situations, high quality decisions are required.

Decision-making is an important task in disaster management because it appears in all management activities before, during and after a catastrophic event [3]. In disaster situations, ineffective decisions can lead to great losses (infrastructure destruction, human losses, etc.). Therefore, decision-making represents an activity that aims to reduce the aftermath of a disaster.

Decision-making process presents a big challenge for disaster stakeholders. Indeed, there are many reasons that make decision making difficult in complex systems such as: interconnected systems, multiple decision makers, multiple

© Springer International Publishing AG 2017
I.M. Dokas et al. (Eds.): ISCRAM-med 2017, LNBIP 301, pp. 122–128, 2017.
DOI: 10.1007/978-3-319-67633-3_10

objectives and constraints, system dynamics and uncertainties in the environment [4].

Crisis management includes various concepts representing different knowledge. To exploit this knowledge in crisis planning and responding, we need to formalize it by using metamodels. The crisis management process, as a collaborative situation, concerns multiple partners [5]. That is why the first step to enhance coordination and collaboration of the crisis stakeholders is representing these partners and the relationships between them in different levels (operational, tactical and strategic) using metamodeling. In addition, we need to model this domain to understand its complexity.

To build a metamodel, we need to identify all generic concepts that appear in crisis management domain [1] then experts in the field should validated it. In our context, a useful metamodel is a metamodel which can:

- Provide a knowledge needed to build a management process in crisis situation, which contains the description of the activities sequence, the involved actors, their roles and the coordination between them.
- Support the decision makers by presenting the information needed to make a decisions.

Consequently, the objective of the present study is to assess whether the existing crisis management metamodels contribute effectively to support decision support processes in Crisis Management. To answer this question, we begin by a deep study of existing metamodels and classify them with regards to the objectives of the current paper.

This article is structured according to the following sections: Sect. 2 presents description of existing crisis management metamodels; Sect. 3 describes a comparative study of the metamodels; finally, Sect. 4 concludes and gives some requirements for an effective metamodel useful to support decision processes in crisis management.

2   Overview of the Metamodels for Crisis Management

Among our objectives in this paper is the study of the existing metamodels that cover the largest number of concepts in the crisis management domain and that present elements in relation to decision-making. Therefore, based on these criteria, we select the most relevant works.

We cannot consider these metamodels as a complete representation of the crisis management domain because each metamodel describes this domain from a specific point of view. Moreover, some metamodels are built to present a specific context or a particular type of crisis.

In this section, we present a brief description of the selected metamodels (Table 1). To describe these metamodels we focused on:

- The aim of the metamodel
- The concepts described by the metamodel
- The formalism description

Table 1. Crisis management metamodels description.

Metamodel	Aims	What does it describe?	How is it described?
[1]	– Lead to better knowledge sharing – Facilitate combining and matching different disaster management activities to best manage the crisis on hand	It contains the relationships among concepts, which can represent the semantic of disaster management domain. It represents the four phases of disaster management (prevention, preparation, response and recovery)	UML diagram based on the four-layer metamodelling framework of the Meta Object Facility (MOF)
[5]	Gather data and generate information usable to design collaborative schema for crisis management (preparation and response)	It describes different crisis management concepts and the relationships among them: – Crisis concepts – Behavior or collaboration process concepts (activities and events to manage the crisis situation) – Context concepts (elements impacted by the crisis) – Collaborative situation concepts (core) – Objectives (actual facts and risks)	Designed as a UML diagram
[6]	– Model for capturing human behavior during flood emergency evacuation – To assess the effectiveness of different flood emergency management procedures	– Human behavior during evacuation in response to a disaster warning – The human decision making process for evacuation – The factors that play an important role in the evacuation decision-making process	– Using system dynamics approach – Causal diagram
[7]	– Present a conceptual model of disasters affecting critical Infrastructures – Deal with communication and coordination among Humans or intelligent agents of the various critical infrastructures	– Concepts to describe a region and the people that occupy it and their well-being – Concepts to describe the various infrastructures that serve this region – Concepts to describe events such as a disaster and its impact on people, directly or indirectly through the infrastructures – Concepts to describe communication and coordination between infrastructures, the regions and people	UML diagram
[8]	Present an organization model for particular crisis (snowstorm)	It represents 3 models: – Snowstorm environmental model: describe the crisis environment – Snowstorm emergency role model: the roles involved in the considered crisis management system – Snowstorm emergency organizational structure and interactions model: the organizational structure of the crisis management system and interactions among actors	– UML diagram – Interaction model

3   Comparative Study

In order to understand the decision processes, their functions, their characteristics, their components and their actors in crisis management and to improve them, we compare the studied metamodels based on different criteria in relation with our objectives. We aim to identify which among these metamodels provides a holistic vision to the studied domain and which among them presents the decision-making elements such as the decision makers and their roles.

3.1   Criteria Definition

Based on the systemic approach [9], we identified some criteria to compare the studied metamodels. The systemic approach, based on the system theory, is often translated, as "The whole is more than the sum of the parts" [9]. It considers the system as a set of interrelated sub-systems. We use this approach because it allows us to understand the complexity of a system. In our context, the crisis and its management are qualified as a complex system [10]. Moreover, to study decision-making in crisis management effectively we should consider the decisional process in the different levels (strategic, tactic and operational) and the multiple actors involved in this process. Therefore, to classify the metamodels presented in Sect. 2, we consider the following questions:

– What aspect(s) presents the metamodel: structural, functional, genetic or teleological?
– What level(s) concerns the metamodel: strategic, tactic and operational?
– Does the metamodel presents the different stakeholders, their roles and the relationships between them or not?
– Does the metamodel concerns particular crisis types or not?

In addition, to model a system from a global view, we need to present it according to these four aspects [9]:

• **Structural aspect:** How the system is composed? We focuses on the relationships between the elements of the system more than the separated subsystems.
• **Functional aspect:** we try to answer the questions: what does the system do in its environment? What is it used for?
• **Genetic aspect:** How the system evolves?
• **Teleological aspect:** What is the purpose and motivation of the system?

3.2   Similarities and Differences

All the metamodels studied in this paper are structural because they contain relationships among concepts (components) which refer to parts of the crisis management domain. Each metamodel shows a particular part of the domain that is why they do not have the same elements. Moreover, they are all functional and teleological because they are built to provide solutions for different crisis activities. For example, the metamodel in [5] is developed to deal with the collaboration problems by modeling the concepts dedicated to describe the collaborative behavior.

The metamodel presented in [6] is dynamic (genetic), it models the human behavior during flood emergency evacuation using causal loop diagram. The metamodel in [8] is the only one that presents the different actors and the relationships among them in the different decisional levels. It proposes a snowstorm emergency organizational structure and describes the communication between actors using an interaction model. These two metamodels [6,8] are focused on some particular types of crisis (the flood and the snowstorm).

The research works in [1,5] are the most relevant and mature until now. In fact, the metamodel presented in [1] is composed of four packages (prevention, preparation, response and recovery). It shows a unified view of common concepts and actions that apply in different crisis. It could be considered as a dynamic metamodel because it models the four phases of crisis management, so, its users can follow the process from the first phase (prevention) to the final phase (recovery).

The metamodel in [5] based on [10,11] is composed of five packages (core, context, partners, objectives and behavior). The main issue treated in this work concerns the coordination and collaboration of the crisis stakeholders. The package "partners" expresses the actors involved in the crisis management without presenting the different actors types (crisis managers, local and state government, crisis service teams) and the relationships between them.

In Table 2, we summarize these metamodels according to the criteria presented in Sect. 3.1:

Table 2. Crisis management metamodels classification.

Metamodel	Aspect				Decision level			Multi-actor?	Particular to some types of crisis?
	Structural	Functional	Genetic	Teleological	Operational	Tactic	Strategic		
[1]	✓	✓	✓	✓	✓			No	No
[5]	✓	✓		✓	✓			No	No
[6]	✓	✓	✓	✓	✓			No	Yes
[7]	✓	✓		✓	✓			No	No
[8]	✓	✓		✓	✓	✓	✓	Yes	Yes

3.3 Discussion: Strengths and Limitations

The studied metamodels are relevant and important. In fact, they all present some advantages. For example, the work in [1] offers the possibility to generate several models specific to crisis prevention, risks management, crisis response, etc. and, based on information stored in a "Disaster Management knowledge Repository" [1], the users of this metamodel could make better decisions. So, it can be used to support decision. Furthermore, the metamodel in [5] is dedicated to support the use of data in crisis management, it can help to insert an "information layer" (interpretation layer) between "data layer" (data gathering) and "knowledge layer" (exploitation layer) [5].

However, all these metamodels have some drawbacks:

- They are dedicated to some precise type of crisis management.
- They do not represent the different types of stakeholders, their roles and the interactions between them.
- They are specifically dedicated to the operational level. They do not represent the other levels (tactic and strategic).
- They present the information needed for making decisions but they do not show the decision making process in crisis management.

4 Conclusion

In this paper, we present the description and the classification of existing crisis management metamodels according to different criteria (structural aspect, functional aspect, etc.). We aim to identify which of these metamodels present a global view of crisis management domain and describe concepts with relations to decision-making. We find that all of them focus on specific point of view and they all dedicated to support decision by providing information that can be needed to make decisions. Some metamodels could be the basis to build a holistic crisis management metamodel (such as the works in [1,5]) because they present a large number of domain concepts, but we believe that they do not offer a sufficient elements concerning the decision making process like presenting the actors, their roles and the interactions among them.

Consequently, they do not support effectively the crisis management process by eliminating the part of the decision-making. So, we can identify some requirements for an effective model useful to support decision processes in crisis management:

- The metamodel should present different aspects (structural, functional, genetic and teleological) to give a holistic vision to the studied domain.
- It should present the concepts related to the decision making process.
- It should show the different actors structures in different decision levels (operational, tactic and strategic).
- It should be generic and not particular to some types of crisis.
- To enhance the decision making process in crisis management, we, preferably, start by the representation of the stakeholders (decision makers), their roles and their relationships.
- The metamodel should be instantiable, useable and practical.

In the future work, we will try, based on the existing metamodels, the crisis reports and others sources, to:

- Identify the different crisis stakeholders, their roles and the relationships among them (For example: identifying the command chain between actors).

– Generate first model representing the crisis actor's structure and the data flow between them. Based on this model, we can generate BPMN (Business Process Management Notation) processes (scenarios) to manage the crisis situation.
– Conceptualize the decision types in crisis management and associate them to the corresponding decision makers.

References

1. Othman, S.H., Beydoun, G.: Model-driven disaster management. Inf. Manage. **50**(5), 218–228 (2013)
2. Asghar, S., Alahakoon, D., Churilov, L.: A Comprehensive Conceptual Model for Disaster Management, pp. 1–15. University of Bradford, UK, and the Feinstein International Center (2006)
3. Cosgrave, J.: Decision making in emergencies. Disaster Prevention Manage. **5**(4), 28–35 (1996)
4. Sarma, V.V.S.: Decision making in complex systems. Syst. Pract. **7**(4), 399–407 (1994)
5. Benaben, F., Lauras, M., Truptil, S., Salatgé, N.: A metamodel for knowledge management in crisis management. In: 49th Hawaii International Conference on System Sciences (2016)
6. Slobodan, S., Sajjad, A.: Computer-based model for flood evacuation emergency planning. Nat. Hazards **34**(1), 25–51 (2005)
7. Kruchten, P., Monu, C.W.K., Sotoodeh, M.: A conceptual model of disasters encompassing multiple stakeholder domains. Int. J. Emerg. Manage. **5**(1), 25–56 (2008)
8. Chaawa, M., Thabet, I., Hanachi, C., Ben Said, L.: Modelling and simulating a crisis management system: an organisational perspective. Enterpr. Inf. Syst. **11**, 534–550 (2016)
9. Le Moigne, J.-L.: La modélisation des systèmes complexes. Dunod Ed, France (1990)
10. Benaben, F., Hanachi, C., Lauras, M., Couget, P., Chapurlat, V.: A metamodel and its ontology to guide crises characterization and its collaborative management. In: Proceedings of the 5th International Conference on Information Systems for Crisis Response and Management (ISCRAM), Washington, DC, USA pp. 189–196 (2008)
11. Lauras, M., Benaben, F., Truptil, S., Lamothe, J., Macé-Ramète, G., Montarnal, A.: A meta-ontology for knowledge acquisition and exploitation of collaborative social systems. In: Proceedings of International Conference on Behavior, Economic and Social Computing, Shanghai (2014)

Serious Games and Simulations

A Crisis Management Serious Game for Responding Extreme Weather Event

Jaziar Radianti[1(✉)], Mattias N. Tronslien[2], Max Emil Moland[2],
Christian A. Kulmus[2], and Kristoffer K. Thomassen[2]

[1] Centre for Integrated Emergency Management, Department of ICT,
University of Agder, Grimstad, Norway
jaziar.radianti@uia.no
[2] Westerdals Oslo School of Arts, Communication and Technology,
Oslo, Norway
me@mattiastronslien.com, max.moland89@gmail.com,
kerutomasu@gmail.com, exlr8.ck@gmail.com

Abstract. Managing crisis with limited resources through a serious game is deemed as one of the ways of training and can be regarded as an alternative to the table-top exercise. This paper presents the so-called "Operasjon Tyrsdal" serious game, inspired by a real case of extreme weather that hit the west coast of Norway. This reference case is used to add realism to the game. The game is designed for a single player, while the mechanics are framed in such way that the player should deal with limited resources, and elevated event pressure over time to manage. Beside applying an iterative Scrum method with seven Sprint cycles, we combined the development work with desk research and involvement of testers, including crisis responders. The resulted game has expected features and behaviors, game(ful), but allow the player to learn through "After Action" report that logs all player's decisions, intended for trigger discussions.

Keywords: Game development · Extreme weather · Resource management game · Serious game · 3D game · Training · Crisis management

1 Introduction

Nowadays, scientific literature in the emergency management domain suggests serious games as a new opportunity of training for the emergency practitioner, volunteers and the public at large is increasing and gaining popularity [1]. For example, a quick search of literature indexed in the Scopus database using exact phrases "serious games" AND "emergency management" between 2005–2017, returned 51 documents. Only five documents or ten percent were published before 2010, and the rest were issued after 2010. Ninety percent of them mention training as a purpose of the games. Why is the serious game (SG) approach for emergency management training becoming popular? The advantages of gaming approach are numerous. It is a less costly alternative for training activities that can complement field training or alternative to the table-top exercise, and allow intended trainees (e.g. emergency management personnel and decision makers) to train more often than they otherwise would be able to do in

© Springer International Publishing AG 2017
I.M. Dokas et al. (Eds.): ISCRAM-med 2017, LNBIP 301, pp. 131–146, 2017.
DOI: 10.1007/978-3-319-67633-3_11

field-based exercises. Well-made training games can replicate conditions in various scenarios, allow organizations to record the exercises and logging the actions for later review, debriefing or repetition. Moreover, of course, games are a less dangerous environment to train new personnel rather than exposing them to risk.

This background has motivated us to develop a 3D SG intended for training the decision making when a responder faces series of events occurred in extreme weather, as reported in this paper. The so-called "Operasjon Tyrsdal" SG is built by taking the extreme weather event Synne happened in the west coast of Norway in 2015 as a case. In the National Risk Analysis 2014 on the disasters that affect Norwegian society [2], extreme weather has been listed as one of six natural events that has the highest likelihood to occur and one that has the greatest societal consequences. Therefore, this scenario is selected in this work.

Yet, as in other crisis scenarios, there are general, typical crisis situations where the emergency responders have limited resources, limited time to respond, and often facing simultaneous events that need prompt responses. Sometimes, a decision maker has to make a priority, which event requires immediate actions, and which event can wait for further actions. Given all complexities in the real-world crisis and the principles of making serious games that need to be taken into account, then the main research question in this paper is: *How can we design an SG for training that incorporates elements of the real-world crisis so that the developed SG is learnable, playable and can be used for decision making and managing extreme weather events?*

The SGs have been applied for fire scenario [3, 4], crisis communication [5–10] or city context [7, 11, 12] as well as extreme weather scenarios such as hurricane [13] and floods [14], but the last two papers require a lot of participants to be able to run the SG. The contribution of this paper lies in our efforts on tailoring the real-world information of extreme weather situations and the game design principles into the "Operasjon Tyrsdal" game. Many actors need to collaborate in a crisis. However, to ensure successful development within a five months' timeline of a bachelor thesis, the SG is designed as a single player game acting as an on-scene commander who is responsible for making decisions and managing resources in a crisis.

An iterative Scrum method was used during the development process, comprising seven Sprints which were divided further into smaller technical development tasks. To allow the involvement of users and local responders in providing feedback during the development process, the language used in the game is in Norwegian. However, for reaching out in the future, it is possible to translate all text and instructions in the game into other languages.

The paper is organized into six sections. After this introduction, the next section briefly presents reviews on related works and state of the art concerning serious game. Section 3 elaborates the scenario and the game design, while Sect. 4 reports the results or end-product, i.e., the "Operasjon Tyrsdal" game. Section 5 describes the users' role as testers in the game development process. Section 6 concludes the paper and reveal potential research directions from this work.

2 Related Works

Games approach with a stress on learning, participation, encouragement, and behavior improvement has been applied in many domains, and therefore repurposing games beyond entertainment is not entirely new [15, 16]. The "serious games" term itself is an old concept, migrating from mainly military use into many domains, denoting the games used for non-entertainment purposes [17], or education [18]. In other words, SG does not have entertainment, enjoyment or fun as a primary purpose, and emphasizing the importance of the careful balance of education and play as a core of this game genre [19]. SG has been applied as well for crisis management, where the intertwining between these two elements have been considered [20–24]. Mitgutsch and Alvarado [25] suggest that the purpose of the SG needs to be coherently reflected throughout the formal conceptual design of the game. Otherwise, the system is conflicting and not cohesive.

Suttie et al. [19] propose a framework that shows the relationship between *game mechanics* and *learning mechanics* through six wide categories of "Thinking Skills" i.e. creating, evaluating, analyzing, applying, understanding and retention. Games that are intended for improving "evaluation skills" seem to fit the spirit of the "Operasjon Tyrsdal" game. The suggested *game mechanics* comprises the following properties: action points, assessment, resource management; while the *suggested learning mechanics* include assessment, incentive, motivation and reflect/discuss properties. Bedwell, Pavlas [26] scrutinize SG literature and analyze the link between SG and expected learning outcomes. The most common learning outcomes of SG literature can be grouped into twelve patterns, i.e., declarative knowledge, motivation, application, adaptation, cognitive strategies, psychomotor, compilation, knowledge organization, attitudinal valuing, automaticity set, internalizing values, and receiving/responding phenomenon. The authors argue that this taxonomy will influence the way SG is designed so that effect on learning can progress purposefully. To sum up, there is no consensus in the SG literature what should be included in learning mechanics and game elements. Yet, there are numerous suggestions on making a good game which can be incorporated into an SG design whenever these elements are suitable for the game purpose.

Now, we shift our attention to exploring the current state of the art of crisis management SG. Di Loreto, Mora [27] analyze the current state-of-the-art of 10 simulation and game type of SGs. The paper identifies elements existing in the SG games. These elements are: a mix between predictable and unpredictable elements; problem dissection; making plans; local optimum vs. global optimum; communication and cooperation; roles; time pressure for decision making; links to territory; asymmetric information; quick convergence toward objective as well as coaching and debriefing. The authors also point out missing elements in current state-of-the-art of SG for crisis management, i.e.: (1) connection with an actual territory is lacking (i.e., the game takes place in an abstracted environment); (2) there is virtually no possibility to play with roles; (3) lacking of emphasis of the importance of the debriefing phase, and (4) no support for coaching during the playing sessions. Our Operasjon Tyrsdal SG at least took into account points 1, 3 and 4 in the development process.

Another direction in SG for realistic training is Pandora [28] that creates a rich multimedia training environment where games and multimedia technologies are heavily used in combination with Augmented and Virtual Reality. It is designed as single site training, deployed and distributed. While SIMFOR SG [29, 30] uses a multi-agent simulation and assessment approach of SG players, targeting the management of distributed and heterogeneous information (in nature or source) based on the concept of Evaluation Space. This is basically a multiplayer game training, but having quite advanced architecture by incorporating Intelligent Tutorial System to enable the playful learning. The Non-Player Characters for SIMFOR SG has been thoroughly studied [29, 30]. Praiwattana and El Rhalibi [31] review another trend for crisis management SG which focuses on existing work regarding automated scenario generation techniques, and not so much on the game development itself.

Lastly, the authors are also aware of some commercial training tools and services in the market dedicated to crisis management, security, and safety training; for example, XVR On Scene [32], Virtual Crisis CRISE [33], RescueSIM [34] and MASA [35]. XVR, CRISE, and RescueSIM provide an interactive 3D environment. In XVR, the user can enter buildings, vehicles and experience a change in conditions, for example, smoke, fire, snow, or flood. The tool can be used to build an incident scenario, provide instant feedback, and challenge users with questions through the exercise. All these platforms are aimed at teaching, learning, and training for emergency handling. MASA has two products, i.e., MASA SWORD for military purpose and MASA SYNERGY that also has a simulation server, a game client, and tool for scenario preparation as many more to prepare and train their senior staff and officials for the management of planned and unplanned crisis and disaster response situations. It is worth to mention that some commercial tools also have some limitations for direct use especially concerning different standard procedures or language issues. The research and development of the SG in this paper are tailored to the local needs, standard procedure and language.

As an initial effort, our "Operasjon Tyrsdal" work is fairly straightforward, as it does not involve sophisticated infrastructure, complex algorithms or multiple players as we see in the Pandora or SIMFOR projects. However, we do include complex mechanics and thorough considerations to reach the status we present in this article.

3   Scenario, Requirements and Game Flow

3.1   Scenario

The game was inspired by big storm "Synne" that occurred in Eigersund on the west coast of Norway in December 2015, where continuous heavy rain caused significant evacuation of inhabitants. The flood reached "red level warning" the highest of four flood alerts that are typically used to tell the danger level of a flood event. Some areas nearby faced risks of avalanches and landslides. The main transport infrastructure both motorways and over 40 secondary roads in two counties were closed, and paralyzing the ground and inter-regional traffic.

The flood level supposedly exceeded the level of the 200-year flood, and the municipality officers could not estimate the height of the water level due to loss of communication with the weather stations. In short, escalated crisis over several days such as Synne offers a unique case for other to learn about managing a sequence of tight events over a longer period with limited resources [36]. When transforming this event into a SG, the time and place is fictional, but the plots, stories, and sequence of events were modified from series of events during this storm "Synne." Finally, the role of the player is On Scene Commander.

3.2   Requirements

We used Unity [37] as a popular platform for game creation, and C# as a programming language. We also set up some high-level technical requirements for this "Operasjon Tyrsdal" game, i.e.:

- The game was limited to a single scenario to allows us to focus on creating a detailed environment with enough features to create a realistic crisis scenario.
- The game would not feature multiplayer activity and concentrate on the individual experience.
- The player would be required to make difficult decisions regarding how to use the resources available. As the game was still designed for the limited environment, we used the local language, i.e. Norwegian, as it facilitated the better involvement of testers at the development stage. Therefore, common local crisis vocabularies were used as much as possible in the game text and stories.
- The game would be stand-alone and playable from a laptop so that users can operate the game without internet or network requirements. The game would only be developed with a target platform of a laptop running Windows, but Unity's' portability means that in the future the game could be ported to multiple platforms.

3.3   Game Flow

The game begins slowly to ease the player into the mechanics, and gradually increase in difficulty and intensity as the crisis worsens. Therefore, it is useful to design the game as similar to crisis phases, where the criticality and complexities gradually increase over time, before the situation finally is normalized. The game flow of "Operasjon Tyrsdal" can be summarized as seen in Fig. 1.

Fig. 1. The game flow

Notification Phase: This is where the game begins, the player is introduced to a tutorial that informs them of the mechanics. After the first introductory event, the player is given full control and will be ready for the next phase.

Action phase: This phase of the game is the early parts of the crisis. The storm has struck the coast, and the waterways are starting to flood. The increased water and heavy winds are having an impact on the region; telephone poles are starting to topple, as well as trees. Residents living in some of the more exposed regions should be evacuated by now. This phase is not too demanding and gives the player a sense of the impending danger.

Crisis phase: The situation is at its worst, the player does not have enough resources to solve every problem, and this creates dilemmas and choices that are not easy to make. An example of this would be for the Player to have to choose to use manpower to strengthen the temporary dam, or abandon it and respond to a landslide with severe consequences.

Normalization phase: The worst of the weather is over, and the game finishes.

This is bookmarked by the player deciding when it is safe for people to return to their homes, and when the first responders are not needed in any immediate and dangerous scenarios. When the "round" ends, statistics for the current game are shown together with a timeline of events, opening up for debate with any non-player participant observers.

4   Graphical User Interface, Game Elements, and Results

In this section, we present the results in terms of Graphical User Interface and game elements. The description of the game elements is presented based on an SG framework and how each element of this context is manifested in the game.

4.1   Graphical User Interface (GUI)

The object in the game can be placed into two categories, GUI objects and game world objects. The numbers 1–6 in Fig. 2 depict the main GUI when the game starts.

The Resources (1) contains the police, ambulance services, fire services and volunteers. **The Notification Dashboard (2)** is an area where the player will get a notification on a situation that requires a response. The player can click the new message (gray) and will be lead into the event location. A successful mission will be marked with green in the dashboard, while unsuccessful mission will be marked with red. **The Resource Tracking (3):** The bars tell players which and how many resources have been deployed in different events. Each bar represents one event being handled. The next section will explain in details the game mechanics of "Operasjon Tyrsdal".

The Action Report Button (4) will trigger an action report, which logs all decisions and interactions that players have performed during the gameplay and the player's answers to the quiz. **The Time Counter (5)** shows the game time and the length of the crisis. Compared to the real time, the game-time is condensed. **The Setting button (6)** can be used for regulating the screen resolution.

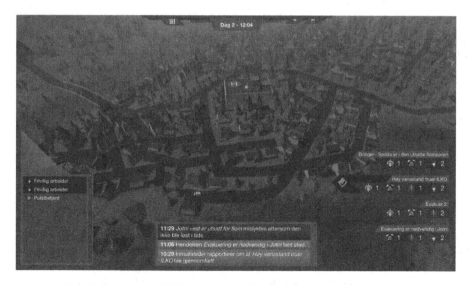

Fig. 2. Overview of GUI when the game is started (Color figure online)

4.2 Game Elements and Results

The framework used for describing the game elements is adapted from Mitgutsch and Alvarado's work [25] who applied it for evaluating SG (Fig. 3). We applied the framework to ensure some essential elements of SG are fulfilled and expressed in the proposed game. This framework offers a structure to study the formal conceptual design in relation to a game's implicit and explicit purposes. The description of the overall "Operasjon Tyrsdal" game will follow the components illustrated in Fig. 3.

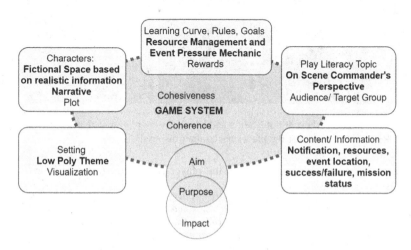

Fig. 3. Serious game design framework, adapted from Mitgutsch and Alvarado (2012)

Game Purpose and Impact: This game is intended as a training tool that allows decision makers (on-scene commander or those who want to learn about decision making at this level) to handle future events in an extreme weather disaster. As a decision maker in the field, the player has to manage the available resources, uphold communication and restore order to the affected area. His/her task includes managing services such as police, firefighters, and medical personnel. The expected impact on the player is the ability to reflect and discuss the decisions that have been made for each pre-defined, automatically triggered event, and what can be done better to encounter similar situations.

Content and Information: As suggested by Mitgutsch and Alvarado [25], the content and information refers to the information and facts used in the SG that could be well presented, adequately formulated, "correct" **or** irrelevant, hard to access or insufficient, and in worst cases, just wrong and biased. In the "Operasjon Tyrsdal" SG, the main information is grounded from the real Synne extreme weather event and timeline [36]. Some information is presented as an alert message to the player, with slightly modifications such as simplification, the sequence of events, anonymizing names, and detailed messages. The information on an event is triggered by clicking the incoming notification in the dashboard in the lower middle of the GUI. Details of an event will appear in a window, in the middle of the GUI (Fig. 4, left). The player is also supplied with the information of their available/accessible resources (Fig. 4, right). The player can see figures of deployed personnel in the information window.

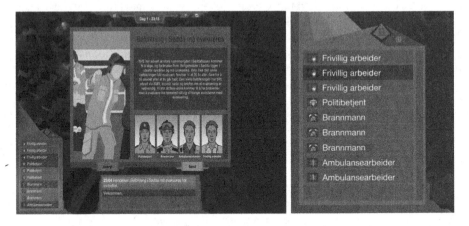

Fig. 4. Left: The detailed event notification and how units are represented and used in the event window. Right: what units that the player has at their disposal:

One example of a message given to the player can be illustrated as follow:

"The NVE (The Norwegian Water Resources and Energy Directorate) has warned that large amounts of water in Søddafossen will rise and cause flooding. Residential areas in Sødda are in vulnerable areas and must be evacuated. Unless the civilian population is evacuated, we anticipate that 20 lives are at risk of being injured or for life to be lost. The civilian population has been warned by SMS, email, radio, and phone that evacuation is necessary. We believe that more elderly people will have trouble evacuating from their homes and will need assistance with evacuation".

The player should think based on some relevant keywords in the message before acting. To support further content realism, we refer to the Police Contingency Guidelines that also explain the roles and responsibilities of the police during incidences such as floods, forest fires, and landslides.

Play and Target Group: The gameplay has a strong focus on the player's ability to manage the limited resources at their disposal. Some inspiration for this game is drawn from similar games in the management genre, such as *Operator: 911* [38], *This is the Police* [39], and *EmergeNYC* [40]. A common gameplay element in these games is a more hands-off approach to problem-solving. The reason for choosing this resource management genre is that we decided to train decision making from the perspective of one who is responsible for coordinating the resources in the field, i.e. the police as on-scene commander. The target groups for learning is the newer responders who learn the leadership in the field in a crisis. Outside the target group, the wider audience who want to understand dealing with extreme weather crisis management can receive benefit from this game.

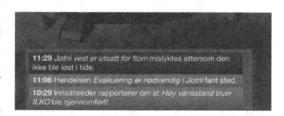

Fig. 5. Alert Notification area

Learning, Goals, and Rewards: When a user plays, the learning points are covered in three different areas: (1) During the play itself, the user should understand the notification and report on a situation correctly, to be able to mobilize appropriate personnel. Figure 5 shows how the message will be available to the player to discern. The notification with green color indicates successful mission with the following message: "10:29 On Scene Commander reports that *High level of water that threat ILKO (place for On Scene Commander)* has been completed". The notification with gray color tells the user an incoming or ongoing event: "11:29 *The necessary evacuation in Jotni* is taken place". The top red message indicates that the action was too late with an example of the following message: "11:29 *West Jotni is prone to flooding* failed as it was not resolved on time"; (2) The speed of making decisions; as the time evolves and the crisis turns acute, the new event notifications are tighter. The resources can deplete to zero, and a player should understand the correct time when to also include volunteers in the response. (3) A quiz to answer will sometimes pop-up in the middle of the game

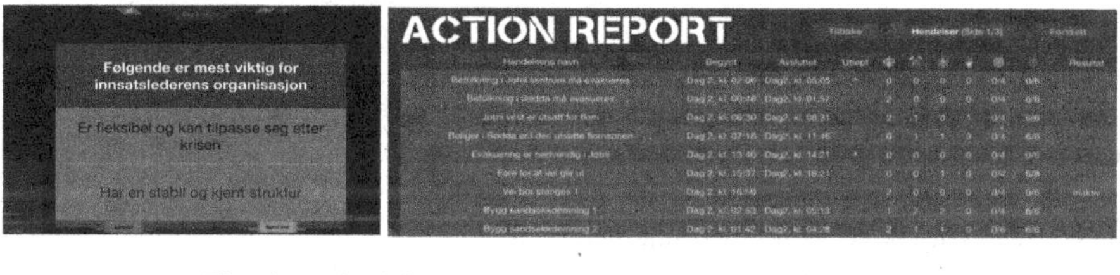

Fig. 6. Quiz (left Figure) and Action Reports (right figure)

(Fig. 6, left). The questions can be true or false, yes or no, and multiple choices. The questions are straightforward and short so that they do not distract the game play. Points are assigned as a reward for right answers to the questions.

(4) Action reports and quiz results can be used for discussion after gameplay. As all decisions for each event and each answer to the quiz are logged, the user can discuss if they have made a correct decision or alternative actions. This action report updates automatically when there is a new entry into the log. All entries are available in chronological order with the most recent being at the top. The contents of the logs are saved in a format that makes it available for analysis after the game ends.

Fig. 7. Left: An arrow indicates the event location (Translated message: Jotni West is prone to flood). Right: Success and failure indicators (Translated message: "High Water level threat place for On-Scene Commander") (Color figure online)

In addition to these learning points, "Operasjon Tyrsdal" will also notify the player of some indicators, such as event location and remaining time available to manage the crisis (Fig. 7 left), and success (green circle) and failure (red circle) indicators. To convey more information to the player, when the user sends units to respond an event, he/she can see information about the event by clicking the dedicated element in the log. If they had forgotten how many units they sent, there would not be any way of finding out.

An event sidebar was designed to alleviate this problem (see Fig. 8). The idea was to have the sidebar become active when the user had dispatched units to an event. The name of the event and dispatched units would be visible to the player for as long as they are not at their station. The player can click on the event name to get more information on the event, or click on the various icons to move the camera to the dispatched units, and see where they are in the game world. When an event completes, an icon signifying if the event was a success or not is displayed.

Fig. 8. Event Sidebar

Characters and Plot: In the game world, the player takes on the role of an "On-Scene Commander" of a small municipality on the west coast of Norway. The player is notified that heavy rains during the past weeks have increased the chances of floods in the municipality and that more rain is expected in the days ahead. It is vital that the player prevents a potential crisis from occurring. The player must manage the resources available in an efficient manner during the extreme weather. During the game, the player will develop a better understanding of how all the different pieces of crisis management work together and what the function of each piece is, and how best to spend those pieces as a resource.

Setting and Visualization: The map visualization and topography setting of the real extreme weather location were transferred into the game world, but the names of the places are fictional, as seen in Fig. 9 left. The visualization also resembles the west coast style of houses with bright colors. Since the environment of Norway's west coast heavily inspires the game, the game area also needs to resemble this atmosphere. Fjords, mountains and rolling hills are core design elements for the game's landscape. The visual graphic of the game is designed with a low-poly theme. This style emphasizes more blocky designs of both models, textures, and environment. It presents heavy winds and rain that lead to floods. The heavy downpour leads to washing the colors and low visibility. It is that cold and rainy atmosphere we wanted to capture in our game, as seen in the Fig. 9 (right).

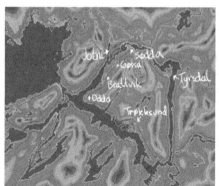

Fig. 9. *Left:* A real map that is transferred into the game world. *Right:* Low-Poly Visualization Style (Color figure online)

Finally, to sum up, we have presented some results tailored in a piece of playable SG with multiple elements interacted coherently and cohesively in the "Operasjon Tyrsdal" game system. Now, we return to our initial research question posed in the introduction section: *How can we design an SG for training that incorporates elements of the real-world crisis so that the developed SG is learnable, playable and can be used for decision making and managing extreme weather events?* The question has been answered through the development of the game world. Designed as a resource management game, it has succeeded to incorporate the real-world events in the game, presents stress and the dilemma of decision-making, quiz, and action point reports to trigger discussions e.g. with other players in a learning group, or with supervisor/trainer in a learning setting as described in this section. Most features in the game have fulfilled the requirements defined prior to the SG development, and these features are functioning well as expected.

5   User Involvement in the Sprint Cycles

Recall that we used a software development *Scrum* method [41, 42] during the "Operasjon Tyrsdal" development, which was divided into 7 *Sprint* cycles within five months. Typically, in a Scrum approach, different people involved in a software development project is divided into different roles: product owner (who maintains product backlog, prioritize product requirements, define features of the product), Scrum Master (who has a leadership role over team members and is responsible for running daily Scrum) and Scrum Team (the developers). In our group, the communications were maintained through different levels of Scrum meetings that addressed some important questions such as: what the team had done, what would be the team's plans until the next meeting, what were the barriers for implementation. We used Trello boards (www.trello.com) for maintaining the artifacts such as product backlog and *Sprint* backlog.

A *Sprint* is basically a set of product implementation cycles that includes the development, the reviewing and adjusting of the software product, and are undertaken during a pre-determined timeframe. In our project, the sprints were designed as a flexible process but were always started with agreed tasks to do. Each cycle was split further into main tasks and sub-tasks. In the first three sprints, involving users were still hard as the main activities focused on finding out the mechanics of the game, how all information could be tailored to the game, and making sure that the mechanics worked well. Thus, visualization was not a priority. Likewise, decent stories and text were still not a focus, nor was user testing. The reason was that at these three initial stages, we already could detect flaws ourselves, and the game was not yet stable enough to show to third parties.

User feedback was elicited only after Sprint 4. We tested the game on six users who had never seen the game before, but we did not yet consider the game ready for testing on actual responders. First-time user perspectives were valuable, to detect if they had a problem with understanding and using the game. We received useful feedback, e.g.:

- the game-control keys were not so intuitive to use;
- the title of each event was not unique so that the user could not differentiate one event from another, and couldn't see if an event was solved or not;
- the game needed an opening that informs the players what to do;
- the meaning of the red and green circles in the evacuation board was not clear;
- the tester did not know what to expect after sending resources, and had no idea how many resources had been deployed and how many resources were still accessible;

In addition to input on these functional features and some minor technical difficulties, we also received comments on some visual design ideas that have been useful to improve the look of the game. All issues were resolved in the following Sprint.

In Sprint 5, a head of local fire service with more than 20 years' experience tested the game. Here we noticed the differences on how he treated the game compared to the normal testers. For example, the user would all the time refer to the game time counter before taking a decision and highlighted how important it was for the player to understand this condensed game time and consequences to the success or failure of the mission. He provided deeper reasons, considerations, and explanation before taking decisions. Upon this understanding, we interpreted that some of the game goals for learning and training were well conveyed. However, he pointed out the need to provide players with some background information, so that they know what to do. This feedback was also handled in the Sprint 6 product.

We had the opportunity to show the game to police personnel at the Sprint 6, however, at this point, we regarded the feedback at this stage as a current limitation of this SG. For example:

- To incorporate data and measures that are needed to make decisions.
- To add more realistic visual graphics such as a crowd of people being evacuated, raised water, and other statistical indicators and big data.

At Sprint 6 and 7, the main development focus was to polish and ensure that all learning points are well received: the action reports, resource trackers, success-failure missions and quiz. From the learning perspective, we deem that the result produced in Sprint 7 shows decent game mechanics, features and learning points, and meets the requirements laid down in the beginning, especially when considering time available for the development. Indeed, more thorough user testing is needed, to know the usefulness of the game as an alternative tool for emergency management training.

6 Conclusion

In this paper, we have presented an SG for training on extreme weather events, where the mechanics follow the resource management genre and have fulfilled the defined requirements. During the development process, we have involved several testers, including fire service personnel and police, especially to test the realism of the game, where the feedback was followed up and expressed in the newer game versions. Innovation in this work is that it is "light" SG that can be deployed anytime, anywhere, running in independent, graphically less-sophisticated computers or laptops. It offers the possibility to conduct debriefing, trigger a discussion with a coach, which is actually missing in the current state-of-the arts. In addition the way we tailored several features have been implemented in the game, such as message notification with success failure code, resource tracking bars, a way to assign resources, information on time available to solve a case, quiz as well as after action report summary, is a part of our contributions to the 3D video-based serious game for emergency management.

Limitations of this work can be listed as follow: (1) We have a one level game as we prioritized that the many complex aspects that are taken into account in the game are represented in the game world. More advanced levels can be another way to go, which can be considered as future work. (2) We believe that the extreme weather case and components in the game are a useful case in a wider context. Thus, technically, it is possible to make it accessible in other languages to reach a wider audience who want to learn about managing resources and time pressure in extreme weather scenarios.

(3) The game was only developed with a target platform of a laptop running Windows, but Unity's' portability means that in the future the game could be ported to multiple platforms. (4) The user testing was conducted in the Sprint sessions. In the future, the usefulness of the game can be evaluated through a thorough evaluation session.

Acknowledgments. The authors are grateful to Kjetil Raeen for the feedback and advice on this work, especially on the game design's point of view.

References

1. Loreto, I.D., Mora, S., Divitini, M.: Collaborative serious games for crisis management: an overview. In: 2012 IEEE 21st International Workshop on Enabling Technologies: Infrastructure for Collaborative Enterprises (2012)
2. DSB, National Risk Analysis, Norwegian Directorate for Civil Protection (DSB) (2014)
3. Radianti, J., Lazreg, M.B., Granmo, O.C.: Fire simulation-based adaptation of SmartRescue app for serious game: design, setup and user experience. Eng. Appl. Artif. Intell. **46**, 312–325 (2015)
4. Wang, B., et al.: BIM based virtual environment for fire emergency evacuation. Sci. World J. **2014**, 22 (2014)
5. Haferkamp, N., et al.: Training disaster communication by means of serious games in virtual environments. Entertainment Comput. **2**(2), 81–88 (2011)
6. Haferkamp, N., Krämer, N.C.: Crisis communication in virtual realities-Evaluation of a serious game on the training of soft skills in crisis management teams. Gruppendynamik und Organisationsberatung **41**(4), 357–373 (2010)
7. Middelhoff, M., et al.: Crowdsourcing and crowdtasking in crisis management lessons learned from a field experiment simulating a flooding in the city of the Hague. In: Proceedings of the 2016 3rd International Conference on Information and Communication Technologies for Disaster Management, ICT-DM 2016 (2017)
8. Ranglund, O.J., et al.: Using games for teaching crisis communication in higher education and training. In: 2016 15th International Conference on Information Technology Based Higher Education and Training, ITHET 2016 (2016)
9. Rocha, R.V., et al.: From behavior modeling to communication, 3D presentation and interaction: an M&S life cycle for serious games for training. In: Proceedings - IEEE International Symposium on Distributed Simulation and Real-Time Applications (2012)
10. Sumi, K., Kasai, K.: A serious game for learning social networking literacy by flaming experiences. In: Poppe, R., Meyer, J.-J., Veltkamp, R., Dastani, M. (eds.) INTETAIN 2016 2016. LNICST, vol. 178, pp. 23–33. Springer, Cham (2017). doi:10.1007/978-3-319-49616-0_3
11. Chambelland, J.C., Gesquière, G.: Complex virtual urban environment modeling from CityGML data and OGC web services: application to the SIMFOR project. In: Proceedings of SPIE - The International Society for Optical Engineering (2012)
12. Lukosch, H., Van Ruijven, T., Verbraeck, A.: The other city - designing a serious game for crisis training in close protection. In: ISCRAM 2012 Conference Proceedings - 9th International Conference on Information Systems for Crisis Response and Management (2012)
13. Brawley, R.S.: Serious games in FEMA Regional Response Coordination Center training and exercises, in Security Studies. 2015, Naval Postgraduate School: Monterey, California

14. Kolen, B., et al.: Evacuation a serious game for preparation. In: 2011 IEEE International Conference on Networking, Sensing and Control (ICNSC). IEEE (2011)
15. Seaborn, K., Fels, D.I.: Gamification in theory and action: A survey. Int. J. Hum. Comput. Stud. **74**, 14–31 (2015)
16. Djaouti, D., et al.: Origins of serious games. In: Ma, M., Oikonomou, A., Jain, L.C. (eds.) Serious Games and Edutainment Applications, pp. 25–43. Springer, London (2011)
17. Metello, M.G., Casanova, M.A., de Carvalho, M.T.M.: Using Serious Game Techniques to Simulate Emergency Situations. in GeoInfo (2008)
18. Linehan, C., et al.: There's no 'I' in 'Emergency Management Team:' designing and evaluating a serious game for training emergency managers in group decision making skills. In: Proceedings of the 39th Conference of the Society for the Advancement of Games & Simulations in Education and Training. Innovation North-Leeds Metropolitan University (2009)
19. Suttie, N., et al.: Introducing the "Serious Games Mechanics", a theoretical framework to analyse relationships between "Game" and "Pedagogical Aspects" of serious games. Procedia Comput. Sci. **15**, 314–315 (2012)
20. Kanat, I.E., et al.: Gamification of emergency response training: a public health example. In: 2013 IEEE International Conference on Intelligence and Security Informatics (ISI) (2013)
21. Dihé, P., et al.: An architecture for integrated crisis management simulation
22. Tsai, M.-H., et al.: Game-based education for disaster prevention. AI Soc. 1–13 (2014)
23. Cremers, A., et al.: Does playing the serious game B-SaFe! make citizens more aware of man-made and natural risks in their environment? J. Risk Res., 1–13 (2014)
24. Matthias, K., Lukas, M., Florian, M.: Research in the large: challenges for large-scale mobile application research- a case study about NFC adoption using gamification via an app store. Int. J. Mob. Hum. Comput. Interact. (IJMHCI) **5**(1), 45–61 (2013)
25. Mitgutsch, K., Alvarado, N.: Purposeful by design?: a serious game design assessment framework. In: Proceedings of the International Conference on the Foundations of Digital Games. ACM, Raleigh, North Carolina, pp. 121–128 (2012)
26. Bedwell, W.L., et al.: Toward a taxonomy linking game attributes to learning. Simul. Gaming **43**(6), 729–760 (2012)
27. Di Loreto, I., Mora, S., Divitini, M.: Collaborative serious games for crisis management: an overview. In: Proceedings of the Workshop on Enabling Technologies: Infrastructure for Collaborative Enterprises, WETICE (2012)
28. MacKinnon, L., Bacon, L.: Developing realistic crisis management training. In: ISCRAM 2012 Conference Proceedings - 9th International Conference on Information Systems for Crisis Response and Management (2012)
29. Oulhaci, A., Tranvouez, E., Fournier, S., Espinasse, B.: Improving players' assessment in crisis management serious games: the SIMFOR project. In: Bellamine Ben Saoud, N., Adam, C., Hanachi, C. (eds.) ISCRAM-med 2015. LNBIP, vol. 233, pp. 85–99. Springer, Cham (2015). doi:10.1007/978-3-319-24399-3_8
30. Oulhaci, M.H.A., et al.: A MultiAgent architecture for collaborative serious game applied to crisis management training: improving adaptability of non player characters. EAI Endorsed Trans. Game-Based Learn. 1 (2014)
31. Praiwattana, P., El Rhalibi, A.: Survey: development and analysis of a games-based crisis scenario generation system. In: El Rhalibi, A., Tian, F., Pan, Z., Liu, B. (eds.) Edutainment 2016. LNCS, vol. 9654, pp. 85–100. Springer, Cham (2016). doi:10.1007/978-3-319-40259-8_8
32. XVR. XVR On Scene. (2017). www.xvr.com. cited 11 June 2017
33. VR-CRISE. Virtual Crisis CRISE (2017). http://www.vr-crisis.com/. cited 11 June 2017

34. RescueSIM. RescueSIM (2017). http://vstepsimulation.com/product/rescuesim/. cited 11 June 2017

35. MASA. MASA SWORD and MASA SYNERGY. https://masa-group.biz/products/synergy/. cited 11 June 2017

36. NTB/The-Local. 200-year flood' ravages southern Norway. https://www.thelocal.no/20151207/200-year-flood-ravages-southern-norway. cited 11 June 2017

37. Unity. Unity, Environment for creating 2D/3D Game that support Multiple Platform (2017). https://unity3d.com/. cited 11 June 2017

38. JutsuGames. 911 Operator (2017). http://jutsugames.com/911/. cited 11 June 2017

39. Weappy. This is the Police. (2017). http://weappy-studio.com/titp/. cited 11 June 2017

40. FlipSwitch. EmergeNYC (2017). https://www.flipswitchgames.net/. cited 11 June 2017

41. Cervone, H.F.: Understanding agile project management methods using Scrum. OCLC Syst. Serv. Int. Digital Libr. Perspect. **27**(1), 18–22 (2011)

42. Schwaber, K.: SCRUM development process. In: Sutherland, J., et al. (eds.) Business Object Design and Implementation: OOPSLA '95 Workshop Proceedings 16 October 1995, Austin, Texas, pp. 117–134. Springer, London (1997)

Learners' Assessment and Evaluation in Serious Games: Approaches and Techniques Review

Ibtissem Daoudi[1,2(✉)], Erwan Tranvouez[2(✉)], Raoudha Chebil[1(✉)], Bernard Espinasse[2], and Wided Lejouad Chaari[1]

[1] ENSI, COSMOS, Manouba University, 2010 Manouba, Tunisie
{ibtissem.daoudi,raoudha.chebil,
wided.chaari}@ensi-uma.tn
[2] Aix Marseille University, CNRS, LSIS UMR 7296, 13397 Marseille, France
{erwan.tranvouez,bernard.espinasse}@lsis.org

Abstract. Recently, there has been growing interest in the use of Serious Games (SG), as they provide a more powerful means of knowledge transfer in almost every application domain especially in the crisis management field. With this increasing adoption of SG, designing novel techniques for learners' assessment and evaluation has become of paramount importance to improve learning results and thus to maintain players' motivation. This paper focuses on the learners' assessment and evaluation in SG. After defining assessment and evaluation, we distinguish two main approaches: implicit and explicit. For each of these approaches, we present some techniques currently used in some existing games. Then we compare these different approaches and techniques. This synthesis is expected to help researchers and games creators working in this area and identifying benefits and limitations of these techniques in order to develop a new comprehensive technique that outperforms all existing ones.

Keywords: Serious Games · Crisis management · Game evaluation · Learners' assessment · State of the art · Taxonomy of techniques

1 Introduction

The rapid development in Information and Communication Technologies has introduced the concept of games designed for a serious purpose other than pure entertainment so-called Serious Games (SG) [1]. The goal of the SG is to make the knowledge and/or competencies acquisition more efficient and attractive than classical learning methods. The growing interest for SG environments, especially for training, has raised new needs in terms of learners' assessment and evaluation [2]. This topic constitutes an important component of any adaptive SG as it maintains relevant information about what went right or wrong during a game session. This information is useful since it is exploited in order to provide to learners the most suitable adaptation according to their profiles and learning objectives/needs.

Crisis management represents a fertile playground for SG because of its *availability* and relative *low cost* (compared to field exercises) and the *variety of situations*

© Springer International Publishing AG 2017
I.M. Dokas et al. (Eds.): ISCRAM-med 2017, LNBIP 301, pp. 147–153, 2017.
DOI: 10.1007/978-3-319-67633-3_12

(industrial accident, forest fires, floods, terrorist attacks...) each involving *multiple roles* (first responders, chain of command, civilian officials...) and *collaborative behaviors* (evacuation, victim salvation, decision process...) [1]. This complexity offers Research & Development opportunities characterized by pluridisciplinary and inter-disciplinary contributions.

SG can range from relatively simple (linear scenario) one-shot development[1] supporting an information campaign (marketing oriented) targeting general public awareness to complex training framework with multi-actors scenario reproducing real crisis management situation for professional (simulation oriented)[2]. Somehow correlated, Crisis Management SG either can be an *ad hoc* software solution to a particular need or developed (generated) with dedicated software development environments [9]. So-called *SG generator* (Game Engine only or domain dedicated Computer Aided Software Environment), include however implicit conceptual limitations on game and learning characteristics (scenario complexity, number of players...) depending on their "target" (3D environment, web game...). Moreover, one "cultural traits" of Crisis Management is the importance of assessing post-crisis what happened, which behaviors where adequate and what went wrong in order to improve procedures and/or training. Such debriefing is also required in virtual training environment and completed by automated assessment. This assessment is more complex in a multi-actors context, multi-skills, and emotion management while keeping the players engaged in the crisis scenario. Above providing assessment capabilities, Crisis Management SGs also require to be evaluated in regards to their training capabilities.

This paper presents a survey of this research issue, describing the main techniques and proposing a taxonomy to better organize them. Section 2 defines differences between assessment and evaluation, distinguishes two approaches of learners' assessment and evaluation, and presents a review of the main techniques used in existing SG related to explicit and implicit approaches. Section 3 compares several SG that have been assessed/ evaluated. Finally, conclusions are drawn and directions for future work are presented.

2 Learners' Assessment and Evaluation Approaches in SG

2.1 Assessment Versus Evaluation

Both assessment and evaluation require (qualitative and/or quantitative) data about learners and utilize (direct and/or indirect) measures to understand and analyze learners' behaviors during a learning session. However, *assessment* is defined as a process of collecting and interpreting data about learners in order to provide them feedbacks on their failures and progress and to make then improvements of their current performances; whereas *evaluation* is the process of making judgments about learners' performances or SG effectiveness based on defined criteria [6].

---

[1] Such as « Mr Travel » on traveler behavior.

[2] Such as CRISE solutions (http://www.vr-crisis.com) .

For more clarity, *assessment* can be described as a "formative" measurement implemented and present throughout the entire learning process for the purpose of diagnosing learners' actions and identifying areas of improvement to increase learning quality. *Evaluation* is a "summative" assessment conducted at the end of a learning process in the purpose to test the overall learners' achievements and to draw judgments about learning quality. So, we can conclude that assessment is concerned with learning process, while evaluation focuses on the product (SG). Figure 1 summarizes the key differences and similarities between assessment and evaluation.

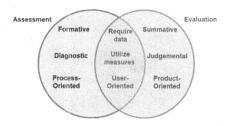

Fig. 1. Assessment Vs Evaluation (adapted from [14])

2.2   Taxonomy of Assessment and Evaluation Techniques

The state of the art of learners' assessment and evaluation in SG is quite rich [6]. In this review, we propose to classify recent existing works into two main approaches according to the technique type used in assessment or evaluation process. The first approach gathers all techniques that assess/evaluate learners explicitly like questionnaires [3, 7, 13] and physiological sensors [10]. The second approach focuses on techniques that assess/evaluate learners implicitly using models and methods of Artificial Intelligence (AI) such as Petri nets and ontology [11] as well as agent technology [2]. The main difference between explicit and implicit techniques relates to the ways of collecting and analyzing data about learners. On the one hand, an *explicit approach* aims to use a direct and obvious measure of collecting and analyzing data about learners. On the other hand, an *implicit approach* aims to collect and to analyze data about learners in an indirect and unobtrusive way, without disrupting the high level of engagement provided by SG. It can be assimilated to stealth assessment [8].

To make the scope of the review more clear, we propose a taxonomy of learners' assessment and evaluation techniques in SG. This taxonomy will structure and guide the survey of SG in the following sections. Figure 2 presents the organization of the key aspects of our taxonomy from the most general to the most specific.

Explicit assessment can be accomplished by using a questionnaire or some sensor devices. In fact, self-report questionnaires [3, 7, 13] are frequently employed because it is simple to implement, but it represents a subjective assessment which relies on non-exhaustive players opinions [6]. Also, questionnaires disrupt the high level of

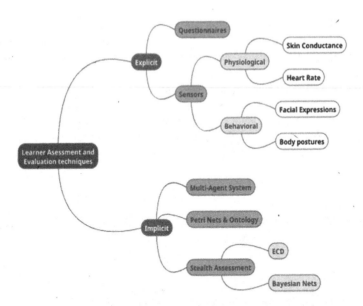

Fig. 2. Taxonomy of learners' assessment and evaluation techniques

engagement provided by SG since they require stopping the learner from playing and requesting her/him to answer questions. Furthermore, the use of hardware and software equipments provides an explicit way to assess/evaluate the learner while using SG [10]. This technique can provide additional information for learner assessment in real-time without stopping him/her during playing. However, it obviously requires the use of additional sophistical devices that can be expensive. In addition, data collected using these equipments can be interpreted in different ways, which can affect negatively the reliability of the learner assessments results [6].

Implicit learners' assessment and evaluation exploits the AI techniques in order to assess/evaluate the behavior of learners such as multi-agent architecture [2], Petri Nets combined with ontology [11] and the conceptual framework Evidence-Centered Design (ECD) combined with Bayesian Nets [8]. All these approaches have the major advantage of adopting implicit models and methods of AI for learners' assessment and evaluation without endangering the high level of engagement provided by SG. Therefore, this type of assessment is intended to support learning and increase learners' motivation. In a Crisis Management context, this may mean "believability" and improves the learning of procedures and best practices. However, most of these approaches consider only one criterion to assess/evaluate learners' reactions to SG adoption in a particular training process.

2.3   Explicit Techniques of Learners' Assessment and Evaluation in SG

Learners' assessment can be performed through evaluating the learners' answers to a questionnaire at the beginning, during or at the end of a game session [3, 7, 13]. For example, Silva et al. [13] invited children, residents of the city of Rio de Janeiro in

Brazil, to use the SG "Stop Disasters" to build a safety culture for emergencies. To assess learners' performances and to verify if the game really improves the awareness of risky situations, the participants answered questionnaires before and after playing the SG about three main aspects namely gameplay, missions and game scenarios.

Advances in neurosciences branch have led to the development of various equipments able to detect and recognize human emotions via facial expressions and physiological signals. Several works have shown that these measures can provide an indication of learners' emotions [10]. For instance, Mora et al. [10] showed the usefulness of collecting data from the WATCHiT sensor during a training event to support debriefing in the crisis management field by addressing two different scenarios. This debriefing, based on sensor data, is considered as a form of evaluation with explicit attention to emotions as well as ideas and behaviors of learners.

2.4 Implicit Techniques of Learners' Assessment and Evaluation in SG

Learners' assessment while playing a serious game can be supported by the agent technology. For example, Oulhaci et al. [2] presented a multi-criteria and distributed assessment approach of learners in "SIMFOR" SG. They propose a methodological framework for learners' assessment based on the concept of Evaluation Space allowing the production of individual and collective multicriteria assessments. In order to implement this methodological assessment framework, they have developed an agent-based architecture improving Non-Player Character (NPC) adaptability (simulation of NPC behavior) and supporting individual and collective learners' assessment. Moreover, Shute [8] proposed an assessment approach embedded within a SG based on the conceptual framework Evidence-Centered Design (ECD) and Bayesian networks in order to model and assess important competencies. In addition, Pradeepa et al. [11] developed an assessment approach that combines a Petri Network and ontology to track not only the player's actions but also to analyze and diagnose the knowledge acquisition of the learner.

3 Comparative Study of SG Assessment and Evaluation

This section describes crisis management SGs providing learner assessment/evaluation during game play. These SG are classified according to the proposed taxonomy. As shown in Table 1, the process of learners' assessment and evaluation in SG requires several inputs collected from the learners' interaction with the game. The techniques described in this article exploit these inputs to extract useful information about the learner(s) and to assess/evaluate his (their) behaviors (outputs).

Table 1 shows that most of SG have been evaluated using explicit techniques. For example, Stop Disasters [13], GDACS mobile [7] and DREAD-ED [3] were evaluated via learners' answers to questionnaires. Table 1 also indicates that only the SG "SIMFOR" [2] used a multi-agent architecture as an implicit technique in order to produce individual and collective assessments. To sum up, we conclude that there is a lack of works considering the emotion concept in learners' assessment in the context of crisis management SG. In fact, human emotions play a huge role in the process of group

Table 1. Serious games and learners' assessment and evaluation techniques

SG	Multi-players	Approach	Inputs	Outputs	Techniques
SimFor [2]	Yes	Implicit Assessment & Evaluation	Learner' Model	Learners' Skills & Social Interactions	Multi-Agent System
DREAD-ED [3]	Yes	Explicit Evaluation	Interview & questionnaire responses	Learners' Social Skills	Interview & Questionnaire
Stop Disasters [13]	No	Explicit Evaluation	Questionnaire responses	Learners' Skills	Questionnaire
SPRITE [5]	No	Implicit Evaluation	Player score	Learners' Skills	Scoring
GDACS mobile [7]	Yes	Explicit Evaluation	Questionnaire responses	Learners' interactions	Debriefing & questionnaire

decision making. In crisis management filed, feeling negative emotions like stress and fear has a negative impact on the individual performance of player during a crisis response. This consequence can affect negatively the collective performance of the group and thus the success of a game session. Additionally, the works exploiting the social interactions aspect in a collaborative context of crisis management games are limited [3, 4]. However, it is important to address the role of social relationships between the different actors in affecting group decision making. In fact, in order to make successful a teamed crisis management, each member should contribute equally and communicate all relevant information to others before making a joint decision.

To tackle this problem, a new technological phenomenon, called "Educational Data Mining" (EDM), proposes to explore big data capabilities in an educational context. It is defined as an emerging discipline concerned with developing, researching and applying computerized methods for exploring data that come from the educational setting and using those methods to better understand learners' behaviors and the settings which they learn in [12]. EDM can be useful in the field of crisis management SG since it manipulates heterogeneous data representing actors' actions, attitudes and interactions relating to a crisis as well as their consequences once the scenario played.

4   Conclusion

This paper presents an overview of Crisis Management SG focused on their learners' assessment and evaluation capabilities. This synthesis can help researchers and game creators by enlighten the main criteria and techniques for learners' assessment and evaluation. The described benefits and limitations of each technique may facilitate the choice of the most adequate way to evaluate a particular SG. Despite the large scope of this survey, this work does not claim to include all existing techniques of learners' assessment and evaluation. However, it includes major themes identified in the literature, and provides a taxonomy where other works can be classified. An important research direction emerging from this research is the development of new implicit assessment/evaluation approaches that consider both emotional and social dimensions in multi-actors SG for crisis management. These approaches can be embedded into SG to provide learners with relevant information about their emotional and social states and to improve training results.

References

1. Walker, W.E., Giddings, J., Armstrong, S.: Training and learning for crisis management using a virtual simulation/gaming environment. Cogn. Technol. Work **13**(3), 163–173 (2011)
2. Oulhaci, A., Tranvouez, E., Fournier, S., Espinasse, B.: Improving players' assessment in crisis management serious games: the SIMFOR project. In: Bellamine Ben Saoud, N., Adam, C., Hanachi, C. (eds.) ISCRAM-med 2015. LNBIP, vol. 233, pp. 85–99. Springer, Cham (2015). doi:10.1007/978-3-319-24399-3_8
3. Haferkamp, N., Kraemer, N.C., Linehan, C., Schembri, M.: Training disaster communication by means of serious games in virtual environments. Entertainment Comput. **2**(2), 81–88 (2011)
4. van, Ruijven, T., Igor, M., de Mark, B.: Multidisciplinary coordination of on-scene command teams in virtual emergency exercises. Int. J. Crit. Infrastruct. Prot. **9**, 13–23 (2015)
5. Carole, A., Franck, T., Etienne, D., Odile, P., Mira, T.: SPRITE - participatory simulation for raising awareness about coastal flood risk on the Oleron Island. In: International Conference on Information Systems for Crisis Response and Management in Mediterranean Countries, pp. 33–46 (2016)
6. Bellotti, F., Kapralos, B., Lee, K., Moreno-Ger, P., Berta, R.: Assessment in and of serious games: an overview. Advances in Human-Computer Interaction (2013)
7. Auferbauer, D., Berg, R.P., Hellingrath, B., Havlik, D., Middelhoff, M., Pielorz, J., Widera, A.: Crowdsourcing and crowdtasking in crisis management: Lessons learned from a field experiment simulating a flooding in the city of the Hague. In: International Conference on Information and Communication Technologies for Disaster Management (2016)
8. Shute, V.J.: Stealth assessment in computer-based games to support learning. Comput. Games Instr. **55**(2), 503–524 (2011)
9. Di Loreto, I., Mora, S., Divitini, M.: Collaborative serious games for crisis management: an overview. In: Proceedings of the IEEE International Workshop on Enabling Technologies: Infrastructure for Collaborative Enterprises, pp. 352–357 (2012)
10. Mora, S., Divitini, M.: Supporting debriefing with sensor data: a reflective approach to crisis training. In: Hanachi, C., Bénaben, F., Charoy, F. (eds.) ISCRAM-med 2014. LNBIP, vol. 196, pp. 71–84. Springer, Cham (2014). doi:10.1007/978-3-319-11818-5_7
11. Pradeepa, T., Jean-Marc, L., Mathieu, M., Amel Y.: How to evaluate competencies in game-based learning systems automatically? In: Proceedings of International Conference on Intelligent Tutoring Systems, pp. 168–173 (2012)
12. Zagorecki, A., Johnson, D.E.A., Ristvej, J.: Data mining and machine learning in the context of disaster and crisis management. Int. J. Emergency Manage. **9**, 351–365 (2013)
13. Silva, V.S.R., Dargains, A.R., Felício, S.P.A.S., Souza, P.R.A., Sampaio, F., Motta, C.L.R., Borges, M.R.S., Gomes, J.O., Carvalho, P.V.R.: Stop disasters: serious games with elementary school students in Rio de Janeiro. In: International Technology, Education and Development Conference, pp. 1648–1659 (2014)
14. Angelo, T., Cross, K.P.: Classroom Assessment Techniques a Handbook for College Teachers. Jossey-Bass A Wiley Imprint, San Francisco (1993)

An Agent-Based Meta-Model for Response Organization Structures

Chahrazed Labba$^{(\boxtimes)}$, Noura Ben Salah, and Narjès Bellamine Ben Saoud

ENSI-RIADI Laboratory, University of Manouba, Manouba, Tunisia
chahrazedlabba@gmail.com, b.salahnoura@gmail.com,
narjes.bellamine.bensaoud@ensi-uma.tn

Abstract. An effective crisis response requires pre-established response structures as well as a predefined reliable command chain. In the literature, multiple meta-models have been elaborated to describe the disaster management domain. However, in regards to the hierarchical response organizations, we did not find any generic model that can be used to identify the involved multidisciplinary agencies scattered on different decision levels. In this paper, we firstly studied the standardized emergency management command chains and organization structures defined in the emergency plans of France, UK and USA. Then, we elaborate a hierarchical and multi-level agent-based meta-model that defines a generic response command chain structure dependent on the type and severity of a disaster. As proof of concept, we instantiate the proposed meta-model for a given disaster within a given country and then implement it as an agent-based simulator.

Keywords: Basic emergency plan · Preparedness · Response organization structures · Multi-agent systems · Rescue process

1 Introduction

Natural and man-made hazards are frequent, unavoidable and can cause irrevocable losses. Even though no emergency plans can prevent death and property destruction, effective emergency plans are required to decrease damages. The purpose of such plans is mainly to incorporate and coordinate all the facilities including resources and personal staff to respond to any emergency. Indeed, identifying the involved emergency management organizations and command chains structure is one of the common objectives listed in different basic emergency plans and guidance [1–3]. In order to enhance emergency responses and coordination, annual costly trainings are carried in different countries. However, such exercises are restricted to small-scale and it cannot be applied for large-scale disasters where hierarchical and multi-level emergency management organizations with different decision echelons (strategic, tactical and operational) are involved. In addition, for effective disaster management and because of the unpredictability of such events, response structures must be in place in advance, ready to

© Springer International Publishing AG 2017
I.M. Dokas et al. (Eds.): ISCRAM-med 2017, LNBIP 301, pp. 154–167, 2017.
DOI: 10.1007/978-3-319-67633-3_13

be activated on short notice, with lines of responsibility clearly delineated and mechanisms for coordination of efforts already established [15]. To overcome the prohibitive expenditure as well as the limits of the training exercises, simulation presents an economical alternative. It allows developing and testing new strategies for rescuing and enhancing coordination and communication between the different decision levels within a response organization. Agent Based Modeling and Simulation are suitable and powerful tools used to simulate different kinds of complex socio-technical systems. They are largely used because they enable intuitive modeling and rapidly simulation of complex situations in any application area. Agent technology is widely used in the emergency management domain such as in [4–6], since it allows the dynamic representation of the stakeholders. However, most of the existing Agent-based Models and simulators for crisis response [4,5,8,9], are used to predict the consequences of a particular course of action, with a particular level of resources, to a particular emergency situation, at a particular location. Indeed, in the literature [18–20], the disaster management domain have been described and modeled through various meta-models, which intend to describe all the concepts and all the management phases within the domain. However, in regards to the response structures, there is no generic model that is elaborated to identify the involved multidisciplinary agencies scattered on the different decision levels (strategic, tactical and operational). To the best of our knowledge, textual descriptions for organizational rules and policies are only available within the emergency plans. In the frame of this work, contrarily to the existing research work, we are interested in analyzing and designing the organization of the main actors including commanders and coordination groups across all the decision levels. Indeed, based on three conventional emergency management command chains structures defined in the emergency plans of France, UK and USA, we propose a meta-model that presents a generic response organization structure. Our proposed meta-model takes into consideration the type as well as the severity of a given disaster. Further, as proof of concept, we instantiate the proposed meta-model for a fire incident within a department in France and then implement it as an agent-based simulator. The reminder of the paper is organized as follows: Sect. 2 presents the related work. Section 3, highlights three existing response organizations and command chains structures. Section 4 is dedicated for presenting the proposed meta-model. And before concluding, Sect. 5 presents a set of experiment results obtained using our implemented agent-based simulator.

2 Related Work

Since training exercises are prohibitively expensive, restricted and inadequate to present real-life situations, simulation is used as an economical solution to cope with large-scale disasters. Agent-based systems are widely used to model and simulate different emergency phases in the aim of testing new strategies for evacuation and enhancing the coordination and communication between the stakeholders. The RoboCup Rescue [5,10] simulation platform was developed after the 1995 Kobe earthquake events. It is used to test different mitigation strategies, during large-scale urban earthquake. Based on a deep literature study, we distinguish three different decision levels within the response organization including strategic, tactical and operational levels (more details in Sects. 4.1, 4.2, 4.3). Each level is composed of a set of actors and centers to facilitate the response process. In the RoboCup Rescue simulation platform, we can categorize two levels of decisions, which are as follow:

- **The tactical level** integrated with the strategic level called also the discrete level includes the fire brigades center, the ambulance center and the police offices. The centers are used to gather information from the scenes, coordinate the actors and manage resources.
- **The operational level** includes the agents presented in the scene such as the firefighters, police officers and ambulance agents.

SimGenis [4] is a generic, interactive, cooperative simulator of complex emergencies. It is an agent-based simulator developed to design new generic and optimal rescue strategies based on a set of inputs including the initial state of victims, the number of rescuers and the method of communication between the stakeholders. SimGenis represents only the implementation of **the operational decision level** in which the actors enact directly in the incident field. However, the strategic and tactical levels do not appear in the proposed agent-based model. In [8,11], the DEFECTO system is presented. It is a multi-agent system dedicated to improve the training of incident commanders by replacing the real firefighters actors by agents. Indeed, based on field observations, the authors highlight some anomalies related to the training within the fire department, which are as follows:

- The limited number of firefighters and the restricted fire representation lead to small-scale training exercises.
- During training the severity as well as the evolution of fire during the time is not taken into consideration.
- The training requires an off duty number of personal, which minimizes the overall number of ready firefighters in case of real disasters in the city.

Regarding the organization structure, both **the tactical level** (discrete) including the incident commander (human) and **operational level** integrating the agents teams are highlighted. ALADDIN [12,13] is a multi-disciplinary project aiming at modelling and designing fully decentralized systems that are more efficient in term of robustness and scalability. The system takes into consideration the **strategic, tactical** as well as the **operational levels** by implementing and testing different machine learning algorithms. However, it considers decentralized hierarchy, whereas real existing response organization structures are usually centralized. To summarize, the existing agent-based simulators for crisis management domain, usually consider a specific disaster in a specific location and focus more on the operational level within the incident field. There is no explicit work dedicated to present and determine the command chains structure and the response organization according to the catastrophe and its geographical extent within a country. Indeed, usually, the basic emergency plans enumerate the different categories of hazards as well as the levels of severity. For example in [17], the Illinois emergency operations plan enumerates five levels of crises severity including Routine business, Minor event, Significant event, Major event and Catastrophic and presents for each event magnitude the corresponding response operations structure (local, regional, national). In the current work, we intend to determine the command chains structure and the response organizations according to the severity of a given crisis, its geographical extent (local, regional, national) knowing that the geographical administrative division depends on the country.

3 Overview of Existing Emergency Management Organization and Command Chains Structure: France, United Kingdom (UK) and United States of America (USA)

In order to decrease the harmful effects of hazards including natural and man-made disasters, Emergency Management Plans (EMP) are developed in many countries around the world. The purpose of such plans is mainly to incorporate and coordinate all the facilities including resources and personal staff to respond to any emergency. Indeed, identifying the involved emergency management organization as well as the command structure is of big importance to reach an efficient emergency management.

In this context, we refer to three different countries including France, UK and USA to study and analyze their emergency response organizations and command chains structure at different levels. Indeed, the referenced countries including France, UK, USA represented in the Fig. 1, define multiple response echelons based on the severity of the hazards. For each country we consider all the response levels: (i) for France National, Zonal, Departmental, (ii) for UK National, Regional, local and (iii) for USA Federal, State, Regional.

3.1 France, ORSEC, [1,14]

The French authorities define the dispositive ORSEC acronym for *Organisation de la Réponse de Sécurité Civile*.[1] ORSEC was, firstly, defined, in 1952, as a departmental plan executed under the authority of the prefect *Préfet*. In 1987, ORSEC evolved to include zonal plans and basic emergency plans that can be used according to the type and the extent of disasters. In 2004, according to the modernization low, ORSEC becomes the unique organization responsible for managing all emergencies. It involves multiple stakeholders enacting under

---

[1] http://www.interieur.gouv.fr/Le-ministere/Securite-civile, consulted 07 April 2017.

a single authority and mobilize many resources through an operational system taking into consideration the severity of the risk and its geographical extent.

Crises are handled at different levels including communal, departmental, Zonal, National and European levels. The level of response is related to the severity and the evolution of a given risk from minor to major events.

In fact, in the case of minor events including mainly road accident and localized single fire with short duration, the mayor occupies the DOS[2] command position. Whereas, in case of a significant risk spreading on more than one commonality, the Departmental Prefect becomes the DOS. If the consequences of a given disaster exceed the capacities of the department, the Zone defense Prefect intervenes in the conduct of the operations and mobilization of additional resources when required. In the case a major disaster strikes more than one zone, the response will be at the national level handled by the Interministerial crisis management center, as well as Operational center for Interministerial management crisis. The Fig. 1 (First box) presents the command chain structure as well as the multi-level organizations involved in the emergency response at all the levels. Each response echelon incorporates different decision levels including strategic, Tactical and Operational.

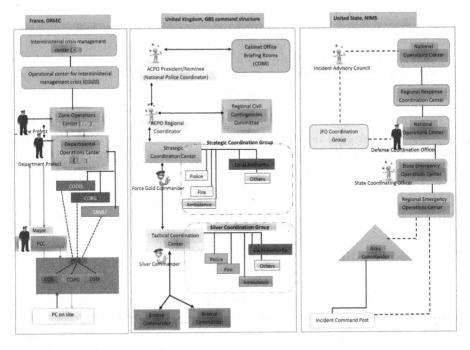

Fig. 1. Emergency response organization and command chain structure respectively in France, UK and USA

---

[2] The DOS acronym for (Directeur d' Opérations de Secours) is a command position depending on the type of the event.

3.2 UK, GBS Command Structure [2]

The UK adopts a national framework of management of response and recovery that defines the role, responsibilities and coordination between the emergency services.[3] The framework incorporates three levels including Bronze, Silver and Gold. Indeed the bronze echelon known also as the operational level is the first to be activated in an emergency response. It integrates first responders including police officers, firefighters, ambulances and specialists advisors. Indeed, within this level the emergency services estimate the extent of the disaster and its severity; assign a particular commander for the affected area in order to ensure coordination between the organizations working in the incident site. Regarding the silver or tactical level, it permits to monitor the actions achieved by the bronze level and ensure more effectiveness in terms of coordination and allocation of resources. The silver commanders are usually located nearby to the incident scene. Finally, there is the Gold or strategic level which is presented by the Strategic Coordinating Group including gold commanders of all the agencies involved in the emergency response. The formed strategic coordination group provides resource support for silver commander and ensures efficient multi-agency response.

Each emergency service contains bronze, silver and gold commanders. The police forces are the leaders in all the levels. However based on the type of the emergencies other agencies may lead the emergency response.

The UK organize meetings in the Cabinet Office Briefing Room A called also COBRA to respond to national and local crises or even to international affairs implicating the UK. COBRA allows the coordination between all the heads of stakeholders involved in the operational level.

3.3 USA, National Incident Command System(NIMS)

In the same context, the United States of America (USA) have the Federal Emergency Management Agency (FEMA) responsible for emergency responses.[4,5] To enhance disasters response, the Homeland Security Presidential Directive-5 was developed in 2003 and incorporates different emergency response agencies. Homeland Security Presidential Directive-5 requires the Secretary of the Department of Homeland Security to develop and administer a National Incident Management System (NIMS), a set of guidelines for use by federal, state, local, and Native American tribal governments, elements of the private sector, and non-governmental organizations to improve their ability to work together [15].

NIMS is defined by FEMA as a systematic, proactive approach to guide departments and agencies at all levels of government, nongovernmental organizations, and the private sector to work together seamlessly and manage incidents

---

[3] https://www.gov.uk/government/policies/emergency-planning, consulted 07 April 2017.

[4] https://www.fema.gov/, consulted 07 April 2017.

[5] http://www.globalsecurity.org/military/library/report/call/call_11-07-ch3.htm, consulted 07 April 2017.

involving all threats and hazards regardless of cause, size, location, or complexity in order to reduce loss of life, property and harm to the environment (See footnote 4).

NIMS incorporates six components including the Incident Command System (ICS) defined as a standardized emergency management system developed in 1970 after the catastrophic fires in California. The ICS combines multiple elements including personals, equipment, procedures and communications between the involved actors in an emergency response. It is represented as a highly hierarchical structure integrating five major functional areas: command, operations, planning, logistics, and finance and administration. In spite of the difference between ORSEC, GSB command structure, and NIMS the three systems defines a multi-agency hierarchical structure including three different echelons: strategic, tactical and operational levels for each administrative division.

However, those systems define a formal centric architecture that cannot be adequate in case of large-scale disasters. In fact, ICS for example, failed to coordinate the different agencies involved in the terrorist attacks in New York City in September 11, 2001 [24]. Indeed, the destruction of the cell phone towers had forced the involved agencies to take separate decisions without having the ability to communicate and coordinate their actions.

In ORSEC, GBS and NIMS the structure of the emergency management organization depends on the event magnitude (severity) to correspond a formal hierarchical structure.

For each response level (national, regional, local), the corresponding response organization is structured into three decision levels including strategic, tactical and operational.

4   Proposed Agent-Based Meta-Model for Generic Response Organization

The proposed Meta-model shown in Fig. 2 organizes the stakeholders and centers into three decision levels including strategic, tactical and operational.

We differentiate two types of agents, the actor and the environment. Indeed the actor represents the different types of stakeholders (police, firefighter, doctors, etc.). The stakeholder can be a commander or a first responder. The environment, within each country, is geographically divided into different administrative divisions (departments, regions, etc.) and embeds the stakeholders as well as the different coordination centers. According to the policies and authorities within a country, a decision level may incorporate one or more coordination centers. Each coordination center integrates multiple representatives of the emergency services involved in the emergency response.

The topology of the response organizational structure depends on the type and severity of a given incident. In [21] five severity scopes are defined including scope I (for small disasters), scope II (for medium disasters), scope III (for large disasters), scope IV (for enormous disasters) and scope V (for Gargantuan disasters). These scopes are defined based on the number of killed persons as well

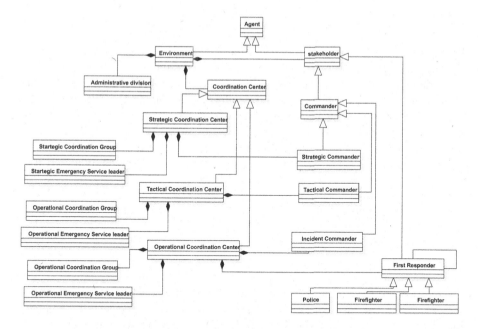

Fig. 2. Agent-based meta-model for generic response organization

as the area of the impacted geographical zone (km^2). Thus the definitions of the response organization is strongly related to the type as well as the severity of the disaster. For example, a small incident requires local efforts (few stakeholders) and few resources to be handled, whereas for a large disaster all the decision levels are activated and inter-regional mutual resources are provided.

4.1 Strategic Level

The strategic level incorporates the strategic commander as well as one or more strategical centers according to the severity of the disaster and the basic emergency plan within a country. It is dedicated to make decisions, provides the inferior levels with the required resources and personal staff during an emergency response and ensures the coordination between the different involved agencies. The strategic commander is responsible for the strategic management of the emergency response and has the following responsibilities: (i) Ensures the activation of all the required emergency response services; (ii) Ensures multi-agencies coordination; (iii) Mobilization of the required resources; (iv) Cooperates with emergency services members to make decisions that are more efficient to meet the requirement of the on-field actors; etc. The responsibilities of the strategic commander may expand or shrink based on the severity of the disaster, the basic emergency plan and law within a given country. Usually the strategic manager enacts within a Strategic Coordination Center (SCC) including supporting teams composed of representatives from the seniors of emergency services involved in the emergency response. The center is located far away from the incident scene.

4.2 Tactical Level

The tactical level includes tactical commander enacting within a Tactical coordinating Group in order to implement the actions and decisions taken in the strategic level. In addition, through this level, the responsible commander coordinates the involved services and shares information among the different levels. He/she is responsible for the tactical management of the emergency response and has the following responsibilities: (i) Cooperates with leaders of the emergency response services on the incident scene; (ii) Determines the suitable way to implement the actions and decisions taken by the strategic commander; (iii) Shares information with superior authorities, etc. Usually the tactical manager enacts within a Tactical Coordination Center (TCC) including the leaders of all the emergency services involved in the incident scene. The center is located nearby the incident scene.

4.3 Operational Level

The operational level incorporates the incident commander and the first responders located at the incident scene and responsible for the execution of the plans and actions elaborated in the superior levels. Contrarily to the strategic and tactical commanders, the incident commander is located at the incident scene. He/she has the following main responsibilities: (i) Coordinates all the response emergency service units on the incident field; (ii) Responsible for the safety of all the stakeholders located on the incident site; (iii) Share Information with superior authorities, etc. The Incident commander works with a support team incorporating members of the emergency services (police, firefighter, doctor, etc) in order to coordinate the actions and execute the actions transmitted by the tactical commander.

5 Case Study

As a case study, we instantiate our agent-based meta-model to simulate the response structure for a medium fire incident (scope II) within a departmental administrative division in France. The incident is a class A [22] fire situation, where only water is used as an extinguishing device. To model the fire evolutionary states, we resorted to a physical mathematical approach proposed in [23]. The fire evolution depends on three elements including the wind speed, the fuel moisture and the fuel weight. Further, to calculate the estimated required fire flow and water to extinguish the fire, we refer to the NFA (National Fire Academy) formula[6] and the NFPA (National Fire Protection Association) standard[7], respectively.

---

[6] http://cfbt-us.com/wordpress/?p=74, consulted 07 April 2017.
[7] http://www.nfpa.org, consulted 07 April 2017.

5.1   Incident Location: Creuse, France

Creuse is a French department located in the region called "Nouvelle-Aquitaine". The fire and rescue departmental service (SDIS[8]) in France incorporates rescue centers at communal level. A rescue center (RC) is responsible for the establishment and readiness of emergency resources including fire engines, firefighters and medical teams. As shown in Fig. 3, Creuse has five rescue centers. At departmental level, CODIS[9] is the operational center that organizes the activity of firefighters, coordinates the different interventions and provides the resources according to the severity of the fire incident. Furthermore, SAMU[10] is a departmental center, which regulates the urgent care resources as well as doctors, nurses and ambulances. It is responsible for the pre-hospital intervention and the coordination of the efforts between the various medical teams within different incident sites in the department.

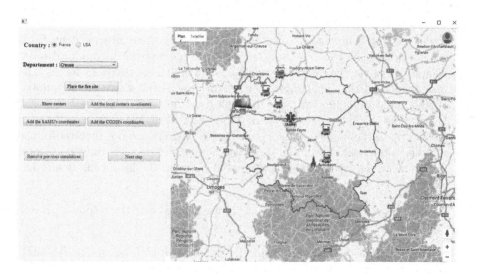

Fig. 3. The different centers used in emergency response

5.2   Experiments and Discussion

The available resources within each center, shown in Fig. 3, are listed in Table 1. The listed resources correspond to both ambulance and fire emergency services. For fires extinction, two types of fire engines with capacity of 3000 L and 8000 L are used. The time required to arrive at the incident site is calculated using the shortest route and without traffic congestion which is defined by Google Maps Distance Matrix API[11]. The following experiments are performed:

---

[8] SDIS acronym for 'Service Départemental d'Incendie et de Secours'.
[9] CODIS acronym for 'Centre Opérationnel Départemental d'Incendie et de Secours'.
[10] SAMU acronym for 'Service d'Aide Médicale d'Urgence'.
[11] https://developers.google.com/maps/documentation/distance-matrix/?hl=fr.

Table 1. The available resources within the centers labeled in Fig. 3

Centers	Number of nurses	Number of doctors	Number of firemen	Number of engines (3000 L)	Number of engines (8000 L)	Time to get to the incident site (Minutes)
RC 1	4	2	6	1	2	12
RC 2	5	3	8	3	2	33
RC 3	3	4	9	2	5	58
RC 4	6	3	6	4	4	61
RC 5	7	6	10	2	4	75
CODIS	—	—	22	9	11	62
SAMU	20	16	—	—	—	37

- Experiment 1: Simulation of a department with three fire incident fields according to *centralized strategy* where we respected the hierarchical organization between the different stakeholders at different decisions levels. The commanders of the departmental centers have a global vision of the number of resources available within the different local centers.
- Experiment 2: Same as experiment 1 but with *distributed strategy*. In such a case, the commanders of the departmental centers are not involved in the response organization. In the case of resource deficiencies, only horizontal communications (reinforcement demands) between the local centers are performed.

In order to assess the impact of the response structure on the response process we consider four initial configurations.

- **Configuration 1:** consists of using (i) the strategy defined in experiment 1, (ii) limited resources and, (iii) high spread rate of fire.
- **Configuration 2:** Same as configuration 1 but with low spread rate of fire.
- **Configuration 3:** Same as configuration 1 but we used the strategy of experiment 2.
- **Configuration 4:** Same as configuration 3 but we used initially low spread rate of fire.

The performed experiments are compared using the following efficiency criteria: (i) the time of pre-hospitalization; (ii) the time to extinguish the fire; and (iii) the geographical area impacted by the incident fire. As shown in both Figs. 4 and 5:

- Configuration 1 represents the second highest values in terms of the predefined efficiency criteria. These results can be explained by the high spread rate of fire, which may lead to a major incident with high level of severity. In such a case, inter-regional response structure needs to be established.

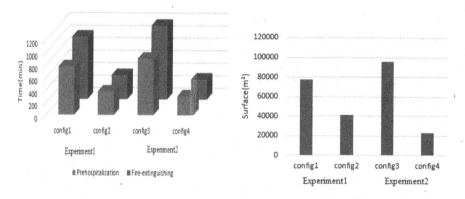

Fig. 4. Pre hospitalization and fire-extinguishing time per configuration for each experiment

Fig. 5. Geographic extend of fire per configuration for each experiment

- Compared to configuration 1, configuration 2 with low spread rate of fire, presents better results. The severity of the incident fire is kept almost constant all the time, which explains the obtained results. Thus the time taken by the response structure to manage the reinforcement demands has a slight impact on the efficiency criteria (reduced values compared to configuration 1).
- Configuration 3 represents the highest values in terms of both pre-hospitalization and fire extinguishing times as well as the impacted geographical area. Thus, the exclusion of the strategic level and the use of a peer-to peer communications within the local rescue centers induces more delays in the response process.
- Compared to configuration 3, configuration 4 presents better results in terms of the pre-defined efficiency criteria. Similarly to configuration 2, the severity of the incident fire is kept almost constant all the time. Thus, within such cases, the response process is likely to be impacted by the dynamic evolution of the disaster over-time more than the response structure. This is confirmed by comparing configuration 2 and configuration 4 where we only vary the used strategies (centralized or decentralized). As shown in Figs. 4 and 5 both configurations 2 and 4 gave closest results.
- configuration 4 provides better results among all the configurations. This can be explained by the fact that in peer-to-peer communications within the same decision level, the resource reinforcement demands may be fulfilled from the first request. Thus, the overheads that are induced by respecting the multi-level hierarchy are not considered in the final response time.

6   Conclusion

In this paper, we presented a hierarchical and multi-level agent-based meta-model that defines a generic response organization structure dependent on the type and severity of a disaster. As proof of concept, we instantiated the proposed meta-model to simulate a fire incident (scope II) within the department Creuse in France. A set of experiments have been performed to highlight the efficiency of establishing the response structure to improve the response time and reduce the impact of the considered disaster.

As a future work, we intend to elaborate coordination strategies between the different decision levels to be used in large-scale disasters.

References

1. Service Interministériel de Défense et de Protection Civile, organisation de la reponse de securite civile, plan orsec departemental-dispositions generales, Avril 2014
2. National Policing Improvement Agency, On behalf of the Association of Chief Police Officers, Guidance on Command and Control (2009)
3. Homeland Security, National Preparedness System (2011)
4. Saoud, N., Ben Mena, T., Dugdale, J., Pavard, B., Ben Ahmed, M.: Assessing large scale emergency rescue plans: an agent-based approach. Int. J. Intell. Control Syst. 11(4), 260–271 (2006)
5. Kleiner, B. Steder, C. Dornhege, D.H.: RoboCupRescue Robot League Team RescueRobots Freiburg (Germany). RoboCup (Osaka, Japan) (2005)
6. Balasubramanian, M., Mehrotra, V.: A simulation framework for emergency response drills. In: Proceedings of ISCRAM, DrillSim (2006)
7. Bharosa, N., Appelman, J., Zanten, B., Zuurmond, A.: Identifying and confirming information and system quality requirements for multi-agency disaster management. In: Proceedings of the 6th International ISCRAM Conference Gothenburg, Sweden, May 2009
8. Marecki, J., Schurr, N., Tambe, M.: Agent-based Simulations for Disaster Rescue Using the DEFACTO Coordination System. In: Emergent Information Technologies and Enabling Policies for Counter Terrorism (2005)
9. Berry, D., Usmani, A.: FireGrid: integrated emergency response and fire safety engineering for the FutureBuilt environment. UK e-Science Programme All Hands Meeting, Nottingham, UK, 19–22 September 2005
10. RoboCup project. http://www.robocup.org/. Consulted 04 Jul 2017
11. Schurr, N., Marecki, J., Tambe, M., Lewis, J.P.: The future of disaster response: humans working with multiagent teams using DEFACTO. In: AAAI Spring Symposium on "AI Technologies for Homeland Security" (2005)
12. Adams, M.F., Gelenbe, E., Hand, D.J.: The aladdin project: intelligent agents for disaster management. In: IARP/EURON Workshop on Robotics for Risky Interventions and Environmental Surveillance (2008)
13. Gianni, D., Loukas, G., Gelenbe, E.: A simulation framework for the investigation of adaptive behaviours in largely populated building evacuation scenarios. In: Organised Adaptation in Multi-Agent Systems workshop at the 7th International Conference on Autonomous Agents and Multiagent Systems, May 2008
14. Organisation de la sécurité civile. http://www.mementodumaire.net/dispositions-generales-2/vigilance-alerte-et-secours/dgv1-organisation-de-la-securite-civile/. Consulted 03 May 2017
15. Kahnl, L.H., Barondess, J.A.: Preparing for disasters: response matrices in USA and UK. J. Urban Health. 85(6): 910–922 (2008). doi:10.1007/s11524-008-9310-y

16. Hawe, G.I., Coates, G., Wilson, D.T., Crouch, R.S.: A survey of usage and implementation. ACM Comput. Surv. 45(1), 51 pages, Article 8, December 2012. http://doi.acm.org/10.1145/2379776.2379784

17. Office of the Governor, executive departments and agencies, the Illinois Terrorism Task Force and the American Red Cross, Illinois emergency operations plan, basic plan and emergency support function annexes, October 2014

18. Othman, S.H., Beydoun, G., Sugumaran, V.: Development and validation of a Disaster Management Metamodel (DMM). Inf. Process. Manag. **50**(2), 235–271 (2014)

19. Othman, S.H., Beydoun, G.: Model-driven disaster management. Inf. Manag. **50**(5), 218–228 (2013)

20. Chen, N.F., et al.: An event metamodel for flood process information management under the sensor web environment. Remote Sens. **7**, 7231–7256 (2015)

21. Gad-El-Hak, M.: The art, science of large-scale disasters. Bull. Polish Acad. Sci. Tech. Sci. **57**(1), 3–34 (2010). Retrieved from 11 June 2017. doi:https://10.2478/v10175-010-0101-8

22. Fire class. http://sp41.free.fr/feux.htm

23. Baines, P.G.: Physical mechanisms for the propagation of surface fires. Math. Comput. Model. **13**(12), 83–94 (1990)

24. Simon, R., Teperman, S.: The world trade center attack: lessons for disaster management. Critical Care. **5**(6), 318–320 (2001)

Towards the Design of Respond Action in Disaster Management Using Knowledge Modeling

Claire Prudhomme[1,2]([✉]), Ana Roxin[2], Christophe Cruz[2], and Frank Boochs[1]

[1] i3mainz University of Applied Sciences,
Lucy-Hillebrand-Str. 2, 55128 Mainz, Germany
`claire.prudhomme@hs-mainz.de`
[2] Le2i FRE2005, CNRS, Arts et Métiers, Univ. Bourgogne Franche-Comté, Bâtiment i3M rue Sully, 21000 Dijon, France

Abstract. This position paper highlights current problems linked to the aspects of the multi-agency collaboration during disaster response. The coordination and cooperation depend on the information sharing and use which must face up to interoperability, access rights, and quality problems. The research project aims at providing an assessment of information impact on the disaster response in order to support the decision-making about what information shared or what quality of data used to improve the response efficiency. Our research approach propose to combine an information system able to integrate heterogeneous data and a simulation system to assess different strategies of information sharing, dissemination and use. A knowledge base is used as a bridge between information system and simulation system. This knowledge base allows for designing dynamically a simulation according to open data and for managing the own knowledge and information known by each agent.

Keywords: Multi-agent based simulation system · semantic · Disaster response · Information management

1 Introduction

The disaster response corresponds to the emergency crisis management and aims at acting to limit injuries, loss of life and damage to property and the environment [1]. This objective cannot be reached efficiently without the collaboration of multi-agency. This collaboration is the main issue of Disaster response. The collaboration has two aspects: the coordination and the cooperation [2]. The coordination is the act of managing interdependencies between activities performed to achieve a goal [3]. The three components of the coordination are activities which correspond to the goal decomposition, actors who are selected to be assigned to an activity, and interdependency which must be managed [3]. The cooperation is defined as "joint pursuit of agreed-on goal(s) in a manner corresponding to a shared understanding about contributions and payoffs" [2].

© Springer International Publishing AG 2017
I.M. Dokas et al. (Eds.): ISCRAM-med 2017, LNBIP 301, pp. 168–174, 2017.
DOI: 10.1007/978-3-319-67633-3_14

Here, the cooperation intervenes in the resource sharing to achieve response activities and in the information sharing to obtain a better view of the crisis situation. Information and communication technologies have a key role to support the collaboration during disaster response by providing solutions of collaboration and cooperation. On the one hand, planning mechanisms allow for organizing actors, activities and resources by taking into account the different interdependencies. On the other hand, the use of semantic technologies through ontologies provides a support for a shared understanding. Several systems have been developed in order to facilitate the collaboration and assess different strategies of collaboration and communication between the different agencies. However, the problem of communication and information sharing intervenes at different levels: Community, Agency and Individual [4]. For the community, the problems concern mainly inter-organizational interdependencies and collaboration procedures where appears a lack of standardization and interoperability due to the need of collaboration between heterogeneous systems. For the level of agencies, the organizational procedures are focused on vertical sharing (means a sharing according to the hierarchical agency structure) of information whose access depends on responsibilities, privacy, security and authentication. For individual level, the execution of a task is influenced by the uncertainty and the time pressure related to the crisis situation. In this situation, each individual is confronted to information access limit, misinterpretation of information due to their quality and the difficulty in determining what information should be shared. [5] claims that "Sharing and dissemination of information is both crucial and problematic". Among the different problems of information sharing, there is a lack of common vocabulary, and a lack of information about the scale of the disaster which conducts to an imprecise information. Moreover, another aspect of the importance of information for the victims is presented. A lack of updated information can increase fear, stress and other emotions of the victims. The sharing and dissemination of information is the heart of the response where the decision-making about the type, the quantity, and the quality of information is crucial. This decision-making during the response is difficult due to the time pressure of the situation. That is why, it is necessary to be prepared in order to make efficient decision about information sharing during the response. Consequently, this paper presents an approach based on a simulation system and a semantic geographic information system to support the preparation of decision-making about information sharing for a disaster situation. Through the combination of a simulation system and an information system, different strategies for the use, the dissemination and the sharing of information can be simulated a disaster response situation. The analysis of simulation results for different strategies aims at identifying the most suitable data to use, the most suitable information to share and disseminate. Section 2 presents works related to the problematic aspects of such a system.

2   Related Work

The problematic to create a system which supports the decision-making preparation about information sharing for a disaster situation, has two problematic aspects. The first aspect concerns the requirements in information sharing and the second corresponds to the evaluation of the different strategies on the disaster response. The two followings section deals with these aspects.

2.1   Information Sharing

Several systems have been developed to facilitate the information sharing during the response. The project Disaster 2.0 offers a web platform where a description of situation needs can be deposed and stored thanks to the use of an ontology. This ontology aims at facilitating the research of specific needs for an agency to bring solutions to the needs corresponding to its capacity [6]. This type of information exchange provides an overview of situation needs and facilitate the management of these response needs. The paper [7] presents the ontology Einsim which aims at providing a simulation model allowing the combination with other response systems. The use of this ontology facilitates the reuse of information coming from response systems. Many common vocabularies have been developed to provide a common understanding and facilitate the cooperation. The management of a crisis (MOAC) vocabulary allows the description of humanitarian activities related to a crisis situation [8]. These activities can be specified by the three questions: Who? What? Where? Emergel ontology [9] has been developed during a European project to facilitate the collaboration between European country through a common language including a specification of pictograms. This ontology provides vocabulary to describe an emergency situation, mainly caused by natural disaster or by the collision of vehicles. Ontologies have been used to provide a common vocabulary allowing a common understanding for the cooperation and facilitating the information exchange. Its advantage is to provide a structured storage of information which can be understood by humans and machines. Ontologies have also a flexibility advantage to combine them or merge them, thanks to matching techniques.

2.2   Strategy Evaluation

The preparation by assessing different strategies aims at determining the most suitable decision. This process is generally computed by a simulation system on different crisis situations using the strategy applications and result assesses for each strategy. The simulator Simgenis has been used to test two organizational strategies (centralized or distributed) and two communication modes (traditional paper or electronic forms) to evaluate their impact on the efficiency of the rescue process [10]. Thanks to this approach, they have identified that the electronic mode of communication reduces delays and victims losses. But, the most interesting is that the most suitable organizational strategy depends on the relation between the number of victims and rescuers. Thus, the assessment of the response efficiency is computed according to the total number of victims, initial state of the victims and the total number of rescuers (e.g. doctors, fire-fighters, and nurses). The Drillsim platform [11] focuses on assessing the impact of information technologies (IT) solution used during the disaster response to show that the use of appropriate IT solutions improve the response. IT solutions are mainly sensors and communication infrastructure including video and audio sensors, people counters, built-in RFID technology, power line, ethernet, and wireless communications. The assessment of the response efficiency depends on the evacuation speed and the number of injuries. The survey [12] presents a framework allowing to assess a response plan and the decision-making of agents by statistical results data according to the crisis incident and its executing plan. The assessment of the response plan can also compute through the use of an ontology and a reasoner to check the consistency of the response plan applied during the simulation [13]. The multi-agent based simulation is the main technology used to test different strategies and assess its impact [10–13]. Among this related work, two main goals are identified. The first is the evaluation of response plans which are mainly applied in a situation of evacuation and only few of them are interested in the aspect of collaboration through organizational strategies or through the consistency of plans. The second is based on the aspect of information, one is focused on the use of information technologies which provide information about the situation and another is focused on communication ways. However, no simulation systems allow for assessing the impact of information sharing and dissemination.

3   Problem Statement

Through the study of the related work and from our knowledge, ontologies for data interoperability and multi-agent based simulation appear as two technologies relevant for the creation of a system able to support decision-making about the information sharing and dissemination. But, we were not able to identify a system that supports the decision-making about information sharing and dissemination for the disaster response. These two technologies must be addressed in the context of disaster response including the following aspects: cooperation, coordination and impact of information. The cooperation requires the management of information through a common vocabulary to facilitate the interoperability, the management of different data sources with different access rights, different quality and provenance. The coordination is influenced by the information, and requires the management of interdependencies between the different tasks, actors and resources in order to respond efficiently. The impact of information on the tasks execution and behavior of victims must be assessed. The information sharing intervenes in the cooperation and impacts the coordination. That is why, the assessment of information impact requires to take into account the cooperation and the coordination aspect. The goal is to show what type of information contributes to an efficient response and how. The simulation of disaster response can illustrate the problems which have been solved thanks to the information sharing. According to the paper [4], this information will reinforce the motivation of response actors for sharing information. It reduces thus, the problem of reluctance to share information with other agencies. Moreover, its authors claim "At the national level, authorities and laws can stimulate inter-organizational information sharing by implementing institutional mechanisms that dictate organizational policies and guidelines" [4]. Thanks to the proof of good practice in the information sharing providing by the simulation system, new laws can be created to improve and stimulate the information sharing during the disaster response. Such a system can generate new decision-making from response actors in the information management and improve the response efficiency. To identify the type of information essential for an efficient response for each response actors (as fire brigades, policemen, etc.), it is necessary to observe their behavior and their requirement during a disaster response. This study, through the use of a multi-agent simulation implies the modelling of these different agents representing the different actors of the response. The current multi-agent based simulations of emergency situation are generally designed for a specific scenario and cannot be reused or adapted to another scenario [7]. This limit, due to a lack of explicit knowledge for the representation of emergency response management, is an obstacle for the reuse and interoperability between simulation models. Our aim is to design a generic model of disaster response simulation. This generic model based on explicit knowledge aims at simulating different situations of disaster and allowing the design adaptation of the simulation. This adaptation depends on the knowledges and information which intervene in its response through diverse actors. Such a model contributes to facilitate the design of response to any disaster by reusability and interoperability of the models.

4   Research Approach

The approach to provide a generic model allowing the simulation of disaster response is based on knowledge modelling. The use of knowledge models aims at providing a flexibility of the simulation system and facilitating the integration and the management of data. These two advantages of the simulation model have an important role for the study of "good practices" in information sharing, which requires the integration and management of information.

Approach. Our approach based on knowledge modelling uses semantic technologies as ontologies to represent the different knowledge necessary for the simulation of disaster response. In literature, some research works use ontologies to model the emergency simulation process [7], and some other use ontologies to know how to use an object perceived by an agent thanks to an environment semantically enrich [14]. However, these two approaches are applied in indoor environments. That is why, they must be extended and adapted to be used in a context of different disaster situations where different actors involve with different goals. The application to a city requires another approach (compared with the application in a building) to obtain a semantic environment. Our approach is to combine our simulation with a semantic geographic information system to provide an environment semantically enrich. The agent model must be able to understand this semantic environment and decide what action is necessary to execute. The goal of these agents is to use its knowledge about how to act and a set of information, provided in input by users to achieve their goal. For example, the knowledge of a fireman is to know what action is adapted to a situation to save people. It is what he has learned to become a fireman. He also knows some information about materials and resources related to its possible actions.

The semantic geographic information system aims at providing this type of information and other information necessary to create the environment by the integration of heterogeneous data. These heterogeneous data are provided by the user to initialize the simulation. These heterogeneous data can come from directly from response actors who have specific data to achieve their tasks, or from open data to create and enrich the simulation components dynamically. A majority of open data is available on Csv files and Shapefile format or other schema-less format, providing a table form which is problematic for a semantic integration. To integrate these types of file, an automatic process based on natural language processing, geographic tools and semantic web has been developed [15]. This work allows for the semantic integration of information for all open data of a city as the different buildings, streets, quarters, fire station, hospitals, city alarm, etc. These types of file provide, first of all, the location, but sometimes also, information as the different services of a hospital and the number of beds for each service. Another example of information from this file is the location of the fire station, the number of professional firemen and the number of voluntary firemen or a building with the number of people inside. All of this information extracted semantically will be used to generate the environment of the simulation and its agent population. The modelling of agents and their interactions will be managed by ontologies about disaster response which will describe the organizational structure of the actors and their interactions.

Plan of Assessment. A use case based on a flood situation in the city of Cologne in Germany is preparing to incorporate plan of the different responder and open data of the city into the knowledge base. Through this use case, a sufficient set of experiments will be led with different input parameters concerning the information sharing and communication. The result of each simulation through the follow-up of variables will be analyzed according to the input parameter in order to identify the impact of information on the response efficiency.

5   Conclusion

The proposed approach aims at supporting the preparation for decision-making about the information sharing and dissemination for a disaster response to improve the response efficiency. The planned approach uses semantic technologies to integrate and manage information necessary for the simulation of disaster response. The management of agents with different knowledge and different information access can allow for assessing the impact of information or its quality on the response efficiency. This impact assessment allows for determining essential information which needs to be shared to obtain a best efficiency.

References

1. Copola, D.: Introduction to International Disaster Management, 2nd edn. Elsevier, Burlington (2011)
2. Gulati, R., Wohlgezogen, F., Zhelyazkov, P.: The two facets of collaboration: cooperation and coordination in strategic alliances. Acad. Manag. Ann. **6**(1), 531–583 (2012)
3. Malone, T.W., Crowston, K.: What is coordination theory and how can it help design cooperative work systems?. In: Proceedings of the 1990 ACM conference on Computer-Supported Cooperative Work, pp. 357–370. ACM, September 1990
4. Bharosa, N., Lee, J., Janssen, M.: Challenges and obstacles in sharing and coordinating information during multi-agency disaster response: propositions from field exercises. Inf. Syst. Front. **12**(1), 49–65 (2010)
5. Manoj, B.S., Baker, A.H.: Communication challenges in emergency response. Commun. ACM **50**(3), 51–53 (2007)
6. Beneito-Montagut, R., Shaw, D., Brewster, C.: Web 2.0 and social media in disaster management: using web 2.0 applications and semantic technologies to strengthen the public resilience to disasters (disaster 2.0 emergency management agencies use and adoption of web 2.0.). Technical report, Aston University, UK (2013)
7. Poveda, G., Serrano, E., Garijo, M.: Designing emergency management services by ontology driven social simulation. IT CoNvergence PRActice **3**(1), 17–32 (2015)
8. Management of a Crisis (MOAC) vocabulary. http://observedchange.com/moac/ns/
9. Emergel. http://vocab.ctic.es/emergel/
10. Saoud, N.B.B., Mena, T.B., Dugdale, J., Pavard, B., Ahmed, M.B.: Assessing large scale emergency rescue plans: an agent based approach. Int. J. Intell. Control Syst. **11**(4), 260–271 (2006)
11. Balasubramanian, V., Massaguer, D., Mehrotra, S., Venkatasubramanian, N.: DrillSim: a simulation framework for emergency response drills. In: Mehrotra, S., Zeng, D.D., Chen, H., Thuraisingham, B., Wang, F.-Y. (eds.) ISI 2006. LNCS, vol. 3975, pp. 237–248. Springer, Heidelberg (2006). doi:10.1007/11760146_21
12. Praiwattana, P., El Rhalibi, A.: Survey: development and analysis of a games-based crisis scenario generation system. In: El Rhalibi, A., Tian, F., Pan, Z., Liu, B. (eds.) Edutainment 2016. LNCS, vol. 9654, pp. 85–100. Springer, Cham (2016). doi:10.1007/978-3-319-40259-8_8
13. Serrano, E., Poveda, G., Garijo, M.: Towards a holistic framework for the evaluation of emergency plans in indoor environments. Sensors **14**(3), 4513–4535 (2014)
14. Durif, T.: Environnement informé sémantiquement enrichi pour la simulation multi-agents: application à la simulation en environnement virtuel 3D (Doctoral dissertation, Dijon) (2014)
15. Prudhomme, C., Homburg, T., Ponciano, J.J., Boochs, F., Roxin, A., Cruz, C.: Automatic integration of spatial data into the semantic web. In: WebIST 2017, April 2017

Collaboration and Information Sharing

Towards a Framework for Cross-Sector Collaboration: Implementing a Resilience Information Portal

Mihoko Sakurai[1(✉)], Tim A. Majchrzak[1], and Vasileios Latinos[2]

[1] CIEM, University of Agder, Kristiansand, Norway
{mihoko.sakurai,timam}@uia.no
[2] ICLEI European Secretariat, Leopoldring 3, 79098 Freiburg, Germany
vasileios.latinos@iclei.org

Abstract. Municipalities play an integral part in the strive for resilient societies. A resilient municipality is not only prepared for short-term shocks such as natural disasters but also more successful in mastering long-term stresses such as profound in-creases or decreases of population. Backed by a large-scale research project we have developed a Resilience Information Portal (RIP). It serves as an artefact to support cross-sector collaboration within a municipality. The core challenges in implementing such a portal reside not in the technological work, though. Rather, it needs to be implemented in the communication and IT strategy of a municipality and be tailored to the processes. In this article, we present key elements for cross-sector collaboration, which were extracted from the implementation process. These insights will prove helpful for municipalities in supporting their journey towards more resilience with technological means.

Keywords: Resilience · Cities · Municipality · Communication · Cross-sector collaboration · IT strategy · Information systems

1 Introduction

Climate change, population growth, natural disasters and the unequal distribution of wealth affect cities in various ways. A municipality, in its role as administrative function of a city, is facing more challenges than ever. Global climate change has been causing heavy rainfall, torrential rain, flooding, and global sea level rises [1]. It is obvious that natural disasters become more frequent [2]. The number of people living in urban areas is increasing [3]. Demographic change is observed in many countries. Not only in the urban areas but globally population is increasing; at the same time, people are suffering social alienation, lack of social cohesion and loneliness [4]. Globalization has dramatically enhanced people's mobility across the nation whereas gap between wealthy and poverty is expanding [5]. In addition, we tend to live longer but maintaining health requires money and other resources.

© Springer International Publishing AG 2017
I.M. Dokas et al. (Eds.): ISCRAM-med 2017, LNBIP 301, pp. 177–192, 2017.
DOI: 10.1007/978-3-319-67633-3_15

Prior study points out that the public sector of a city tends to rely on sectoral approaches and fails to involve relative stakeholders [6]. Cross-sector collaboration is the key for cities to overcome these challenges. Cross-sector collaboration comprises both internal collaboration within a municipality and involvement of external stakeholders whom the presence is essential to approach the problem [7]. This paper addresses how cross-sector collaboration works to enhance urban resilience. Resilience in this paper is defined as

> "to resist, absorb, adapt to and recover from acute shocks and chronic stressed to keep critical services functioning, and to monitor and learn from on-going processes through city and cross-regional collaboration, to increase adaptive abilities and strengthen preparedness by anticipating and appropriately responding to future challenges" [8].

We elaborate on communication challenges that municipalities face in building resilience capabilities and implementation results of the communication platform, named the Resilience Information Portal (RIP) [9]; the portal was developed to tackle these challenges. Implementation results show that key elements for cross-sector collaboration are the alignment of the IT strategy and the communication processes to facilitate an improved collaboration with key stakeholders, as well as the engagement of citizens to resilience building activities. Achieving this alignment is not straightforward, though, and it is not a unidirectional process.

Our work makes two main contributions. First, we report experiences from the work with seven cities on the technological support for resilience-related activities, in particular concerning collaboration, communication, and engagement. Second, we propose steps towards a cross-sectoral collaboration framework in the context of urban resilience.

This paper is structured as follows. In Sect. 2 we describe the background of our work. Detail of the implementation process is shown in Sect. 3. Then results are discussed in Sect. 4, followed by a conclusion in Sect. 5.

2   Background

The Resilience Information Portal was developed under the European Horizon 2020 project, Smart Mature Resilience (SMR), which started in 2015 with four academic partners, a standardization organization, a city consultancy network, and seven European cities as a consortium member. The seven cities belong to different European countries and vary in size, organization, and general characteristics. They are distinguished in the project by tier-1 and tier-2: the former *pilot* the implementation of the project's results while the latter act as evaluation partners.

The project aims at developing a European Resilience Management Guideline for a city until the end of the project period, May 2018. The guideline contains five tools: (1) a maturity model for assessing maturity level of resilience, (2) resilience policies to help a city to enhance resilience, (3) a risk systemicity questionnaire supporting a self-assessment of vulnerabilities, (4) a system dynamics model for simulating achievement toward resilience, and (5) an information portal that promotes stakeholder/citizen engagement and communication.

2.1 Towards a European Resilience Management Guideline

The Resilience Information Portal is one of the main elements of the European Resilience Management Guideline. Moreover, as it provided technological support for communication and collaboration, and thereby ultimately an increased resilience, it is the key of our considerations in this paper.

The tool implementation process for the Resilience Information Portal was organized over a seven month period. Seven cities in consortium were split into two categories; three were selected as a main implementation field, and the remaining four cities participated as peer-reviewing partners.

The implementation activities took place with the support of the local research partners in the category 1 cities, while a city consultancy network was acting as *external coach* and coordinator, facilitating knowledge and information exchange between partners and city officials and representatives. During this period, partners and city stakeholders had the chance to explore and validate the tool in the security sectors that were already identified and to provide input to the tool developers for the finalization of it; input that was used to constantly update the portal's functionalities and improve the tools' qualities.

The main elements of the iterative implementation in each city were:

(1) a 'kick-off workshop' that aimed to gather the most relevant stakeholders in the category 1 cities for the selected security sector, and present the project goals and outputs, introducing participants to the project's resilience management and co-creation approach, and de facto kick-starting the implementation in the city;

(2) a series of *webinars*, together with the peer-reviewing cities that had the opportunity to ask questions and provide their insights and feedback on the ongoing tool development (external stakeholders were always invited to attend the webinars);

(3) a series of informal interviews with all seven cities order to further discuss issues and challenges that could not be touched upon during the webinars, like social media integration into the proposed design principles;

(4) bilateral meetings with identified stakeholders, organized by the city partners to further explore synergies and collaboration potential between institutions, municipal departments and utilities and project consortium; these meetings also aimed to find the most appropriate ways to integrate the Resilience Information Portal with existing communication mechanisms and platforms that city stakeholders have been potentially using;

(5) a review workshop during which the category 1 cities provided feedback on the implementation process to the tool developers, while the category 2 cities shared their additional feedback and summarized their recommendations for the finalization of the tool through a combination of facilitated discussion, based on guiding questions, and conduct of interactive exercises in breakout groups; and

(6) stakeholder training workshops in each category 1 city, once the Resilience Information Portal was already functional in *beta version*; these stakeholder training workshops mainly aimed to present the tool to the most relevant stakeholders and provide with an up-to-date and ready-to-use application that could support the city's resilience building efforts.

The collective feedback from all cities showed that informing thoroughly stakeholders and city representatives is important and necessary in order to secure their active participation and involvement. There is need for further focus on stakeholders that are mostly affected by or interested in an issue or challenge. Especially the stakeholder training workshops were used as a direct knowledge transfer platform that enabled the project partners to take stock of the co-creation activities (see also [10]) that they had been done so far and to understand if this exact creation process was helpful in terms of producing a tool that would be widely applicable in the future and could be transferred to cities that have not followed the co-creation process from scratch. Also, they provided excessive input on how the tool could feed into the Resilience Management Guideline, which is the project's main goal and outcome.

2.2   Communication Challenge and Resilience Information Portal

To identify requirements for the portal functionality, we interviewed six cities from January through April 2016[1]. In total 20 sets of interviews have been conducted, and 33 representatives have participated. Interviewees were from city council, regional government, police, fire brigade, infrastructure maintenance companies, energy suppliers, and research agencies.

Prior to the interview, a pre-questionnaire had been sent to city council in each city to define communication challenges between stakeholders regarding resilience building activities. The pre-questionnaire was delivered online, and contained a small number of questions, which considered communication activities, communication challenges, and relevant stakeholders. All questions were asked in an open fashion and allowed for detailed input from the cities, including additional remarks that would go beyond the questions. We used the results to design and tailor questions for face-to face interviews.

Questions asked in the interviews considered

1. the way of information and knowledge sharing both in an emergency situation and long-term resilience building activities,

---

[1] While it might be guessed that interviewing six out of seven cities had a particular strategic background, it merely derived from the way project work was organized.

2. how participants utilize information systems to enhance communication, and
3. future requirements for an information system.

We coded interview transcriptions and followed the *grounded theory* methodology [11, p. 45],[12, p. 74],[13, p. 274]. For this purpose, we created 18 categories of communication activities. They were further grouped into six core variables, which were treated as design challenges respectively goals. The results show that cities already have information systems to share information within the organization and related stakeholders [14]. However, the overall purpose of those systems is aimed at logging and storing incident information, especially in short-term time span. The systems do not work for knowledge sharing purposes and to enhance cross-sector collaboration to create long-term relationship. Reporting systems are developed through several levels, i.e., national, regional, and local (different departments inside the city council), which results in municipalities operating several different communication tools. This causes information fragmentation and sometimes also a lack of clarity regarding what information is stored where. In addition, each kind of information is shared through different protocols. As discussed in the previous section, cities are facing both comprehensive, complicated shocks and stresses. The number of involved stakeholders increases which causes further information fragmentation and lack of direct communication among cities and stakeholders. As a result, it affects especially cities' responses to the event of emergency. This causes difficulties in collecting related and necessary information as quickly as possible. This can rise complexity even further if additional systems are used to get an overview, but no full integration or at least information update automation is possible.

Communication with stakeholders and citizens to enhance their engagement is a common concern to all cities. Interactive communication between citizens and realization of long-term involvement is desirable. They try out social media to involve mainly citizens. Such efforts are still in the trial phase, though. Cities not yet developed a strategy for utilizing social media. Cities typically lack human resources to cover all tasks surroundings information provision.

Reflecting these findings, we decided to develop a portal toolbox instead of a *single* Resilience Information Portal [9]. It was acknowledged early in the project by the city partners that an off-the-shelf portal would not replace existing IT systems. Rather, a portal would be implemented on-top or in conjunction with existing IT systems. Therefore, cities required guidance and technological help in setting up their portals. Moreover, they were eager to try out functionality within necessarily decided to implement it. Our portal toolbox contains the following basic functions:

- At its core, it in a publicly available Web application following the common standards such as HTML5, and providing the basic functionality known from the Web.
- It must provide possibilities to provide static and dynamic pages.
- Easily updateable content from data sources must be provisioned, such as contact lists.

- Existing systems must be capable of being integrated by several means including *inlinking* and custom display of external data sources.
- Editing must be supported in an easy way that allows non-technical staff to update the portal.
- A user management, as well as a roles and rights management are required.
- There must be the possibility for an *emergency mode*, which can be used in case of disasters, where providing the population very timely with the most important information only is essential.

Regarding its quality, the portal must be very user friendly, adhere to accessibility best practices (and standards, where they exist), and offer a sufficient level of security.

While this functionality is rather ordinary (particularly in the condensed and simplified form presented here), realizing these functions is not so straightforward when they ought to be integrated with many internal and external systems and need to made fit with the corresponding processes in the municipality. We will elaborate on these complexities despite seemingly simple technology in the further course of the paper.

3 Implementing the Resilience Information Portal

The development process of the portal takes a recursive approach, in which we developed a prototype and implemented it in the category 1 cities out of seven, and received feedback from the category 2 cities afterwards. The prototype thereby grows to be a portal toolbox.

3.1 Implementation in the Research Context

Implementation in the context of software engineering is typically used synonymously with programming or development [15]. When we speak of implementation, however, we use it in the sense typically found in the context of business administration. It, thus, does not refer to a technical activity. Rather, implementation denotes the process of preparing, setting up, integrating, and activating structures and processes to an existing system. Thereby, the implementation process takes an abstract idea and concretizes it in the real world. An implementation process may include technological artefacts, but this is no necessity. In the sense of our project, the implementation of tools goes much beyond proposing a functional specification to a city and upon acceptance to program the actual tool in a city-specific form. Rather, much emphasis is given on the early steps of implementation, particularly on preparation and analysis. This caters for the project's approach of exploring novel fields and increasing the resilience of the partner cities *while* learning about the ways of doing so at the same time.

To get more concrete: implementing the portal meant to at first sit together with relevant stakeholders and discuss whether our proposal of a portal (based on cities' input) really captures what *they* consider essential. Then, we jointly explored particularities and idiosyncrasies of the respective city. Depending on how far they are reflected by the intended portal, features where acknowledged, extended, or marked to be of less relevance. Based on this work, we asked cities for exemplary content and build a first test-version of *their* portal for them. This still is not the actual Information Resilience Portal of a city but rather a *playground* for assessing portal functionality and for learning about the possibilities. Therefore, the implementation process is still ongoing and will eventually lead to fully-fledged municipal portal. However, since the already conducted steps are those with the greatest learning and the best possibilities for generalization, we deem the time ripe to report insights from the process – as we have started with this paper.

3.2 Input from Cities

We received a variety of feedbacks from cities throughout all implementation steps introduced in Sect. 2.1. The following three categories are essential to promote cross-sector collaboration: keeping (stakeholders) updated, building a community, and aligning with the communication strategy (i.e. the integration of daily-basis communication and emergency communication). Quotations[2] we use in this section mainly come from the implementation step (6), i.e. stakeholder training workshops. Otherwise the citation source is given right after the quotation.

Keeping Updated. Keeping informed what a city is doing, and which stakeholders are involved, is critical for cross-sector collaboration. One city mentioned: "we are not updated on what the police is doing and I do not think the police is updated on what we are doing" [14]. All cities are facing a *silo* problem which every department has a "myopia lens" and does not get an overview. Another city referred to it this way: "since we are in the same building it is quite easy to meet and communicate, but if we are in different departments, it becomes a challenge to get the correct information to the right people. And this is where we get back to the portal. How to get information there quickly so that everyone involved get it." *This* is how the portal could support a city to make (or improve) collaboration.

Building a Community. All cities agree that trust between the city and the community should be built with the help of the portal operation. The terms *social capital* and *social resilience* are becoming critical factors toward resilience [16–18]. Through the process of trust building, cross-sector collaboration within a municipality is enhanced. It is important for a city to aware of *how* they engage

---

[2] Please note that only minor editing has been done for quotes from oral expressions; we refrained from fixing grammatical or language problems where meaning or connotation was at the risk of being altered.

people. They should understand *who*, *when* and *why* they should be engaged. The portal could work in providing manuals for the public, which guide how to react to a crisis situation. The tool should integrate community mapping in order to engage people of all levels of capability or authority and various disciplines [19]. This mapping should include many aspects like land use, community facilities, evacuation plans, transport options, and so on. Residential associations, volunteers and key personnel from related sectors inside a municipality are forced to work together through creating the community mapping. In addition, the portal fosters a common understanding of challenges which results in raising awareness in a whole city level.

One city mentioned that "the national crises management system is sort of log and reporting system. You cannot add maps or online computer systems into this log to share with other groups who should be able to access this." Such functionality refers the potential of the portal, which connects all relative sectors and stakeholders as a community.

Aligning with the Communication Strategy. Prior studies argue that communication tools that are not used in the daily-basis operation are not useful in an emergency situation [20,21]. A city commented: "we must use the channels (for emergency) in our daily life so that people get used to use them. Accordingly, these channels should have information that is interesting to public from day to day." This applies to the communication structure; collaboration rarely happens between unstructured relationships. Cities need to build formal, informal and non-formal relationships with all relevant and involved stakeholders [19]. These relationships can be supported or enhanced through both analogue dialogue and utilization of information technology. Therefore, the portal should not only target emergency situations, but should instead become an interface of dialogue and exchange between the citizens.

We therefore have to stress the importance of developing a comprehensive communication strategy for resilience, which guides a city to aware people need to be informed. Throughout the implementation process of the portal, several cities stated that the process triggered a lot of discussion within the municipality on existing and needed communication mechanisms [22]. A communication strategy should be aligned to IT management.

Good alignment brings two benefits. First, technologies (the portal in this research) are able to bring together long-term stresses with emergency managers at city level [22]. The degree of alignment affects to outcomes (performance) as a city tends to inform relevant stakeholders with digital communication platforms. The following quotation supports the argument: "When a crisis emerges, we contact the police, and when we get information that it became stable to be published, we do that, and keep updating and share that on municipal web page and social media." Second, as discussed previously, cities already have implemented several information systems to deal with an emergency situation. So this is an important task for cities to prioritize existing tools along with the communication strategy. One question arose by a city is: "We already have many tools that are working well. Sometimes it is easier to fine-tune these tools and make them work better compared to have a new tool and find out that it is already the same. The current way we are using during crisis is functional. So, when it comes to the portal I am not sure if we should adapt another portal, as we already have some stuff working. It is good, but also is it necessary?"

3.3 Intermediate Conclusion

To draw an intermediate conclusion, our observations can be summarizes as depicted in Fig. 1. IT can be seen as the foundation of communication and collaboration activities; IT by itself does not provide additional resilience in the sense of our paper, although it may do so in the context of infrastructure. Communication and collaboration activities both contribute to urban and social resilience. Improved communication has a direct impact in resilience. Collaboration also has a positive influence on resilience, even if non-communicative collaboration activities are concerned. To give an example for the latter, think of sequential data processing by different stakeholders. The data exchange is of course a form of communication, but we distinguish here between personal communication and data exchange, the latter in the figure belonging to the information technology foundation. Particularly collaboration enabled by communication can be seen as the most profound contribution to resilience.

Increasing a municipality's resilience is no one way process. As denoted by the arrows in Fig. 1, building resilience typically takes an approach of setting up better (or more suited) IT systems to enable more or improved communication and collaboration, which in turn foster resilience. This so to speak is a *push* towards resilience, but there is a *pull* as well. When looking at the municipality with an assessment of its current level of resilience, and when building the municipal resilience strategy, an alignment is needed. This alignment must include the IT and the communication strategies, thereby providing what is needed to further increase resilience – though collaboration.

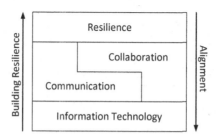

Fig. 1. IT, Communication collaboration and resilience

4 Discussion of Experiences and Findings

From now on, we investigate detail structure between IT, communication, and collaboration. The question is how to mobilize cross-sector collaboration. In this vein, we discuss what elements are essential to the cross-sector collaboration framework. Then, we name open questions, limitations, and future work.

4.1 Key Elements for Cross-Sector Collaboration Framework

The framework for cross-sector collaboration which we ground consists of five components, namely initial conditions, process, structure and governance, contingencies and constraints, and outcomes and accountabilities [23] (see Fig. 2). Initial conditions explain the general environment which affects the formation of collaborations. It contains sector failure that refers to efforts of single-sector solving a problem and failure after the attempt. Direct antecedents mean linking mechanisms which enable collaboration possible. Process components are embedded in collaboration. It overlaps initial conditions and structure. Structure consists of goals, specialization of tasks, rules, standard operating procedures, and designated authority relationships. These structure, governance and process

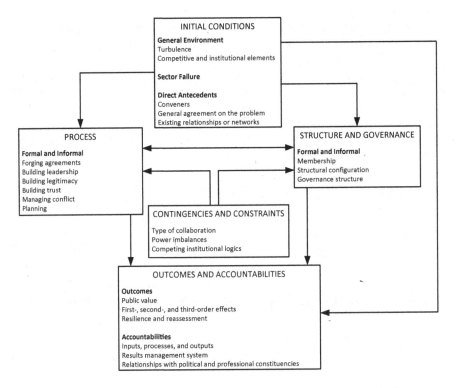

Fig. 2. The original framework for cross-sector collaboration [23]

are influenced by contingencies and constraints. It also affects sustainability of the collaboration. This framework enables us to understand cross-sector collaboration systematically as structured organizational processes. It shows how each component influences each other. Moreover, the framework tells us what is required to enhance collaboration as well as which are the missing parts.

Our extension of the original framework is depicted in Fig. 3. Additions to the original model are depicted in italic font. Items that we do not deem particularly relevant for the perspective that we take have been omitted. Moreover, one main element has been added to the model.

We explain the proposed changes now in detail. First, we have added SETTINGS as an influencing factor. While the initial conditions also only have influence on the other elements of the model but cannot be influenced by them, we deem the initial conditions for collaboration for urban resilience to be twofold. The initial conditions in the sense of the original model are dependent on case-specific factors, such as a municipality's size and existing infrastructure (e.g. regarding the threats) and existing antecedents (such as national structures). These initial conditions underlie the impact of shocks and stresses, which in a way are external to the municipality. While shocks (such as a natural disaster) and stresses (long-term developments, such as migration) are dynamic as they are not fully predictable and change in character and severity of time, they are static to the model. While it would also be possible to include them as abstract threats in the initial conditions and to rename the *threats* to *concrete threats of*

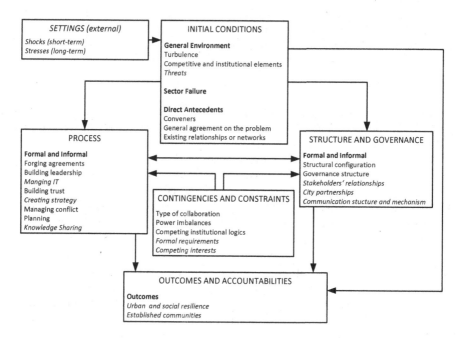

Fig. 3. Our extended framework for cross-sector collaboration, based on [23]

a municipality, we deem the solution of adding an element to the model clearer. Process and structure as well as initial conditions are both critical to influence outcomes.

Regarding the process, *managing the IT* is a key addition. Moreover, *strategies need to be created*, most notably regarding the IT and regarding the municipal communication. The alignment of the communication strategy – which grounds on IT functionalities – is the key elements here. *Knowledge can be shared* through the alignment that results in enhancing urban and social resilience as an outcome; it thereby is the most mature elements as part of the process view.

We also added *communication structure and mechanism* with stakeholders and relative city partners as a new element in the structure component. Establishing such structures helps cities to keep updated to tackle both short-term shocks and long-term stresses.

Municipalities face additional constraints. Laws, regulations and other legal requirements might impede collaboration. Certain actions may not be allowed (e.g. for privacy reasons), security measurements could make communication cumbersome, and documentation needs could render processes to be inefficient. Therefore, we deem formal requirements to be a constraint. Moreover, there can be competing interests. Think e.g. of communicating to the public in a case of disaster. Should the main source of news be the municipality, the policy, or some governmental entity?

Finally, the outcomes have been much simplified in comparison to the original framework. *Urban and social resilience* to a determinable degree is the obvious outcome that is the aim of all activities that our framework describes. On top of that, the cross-sector collaboration will foster *established communities*. These can prove useful beyond a pure resilience scope.

4.2 Open Questions

The proposed framework is a first step towards a better understanding of communication, collaboration, and the dependence of IT when trying to improve resilience. While the current work stresses that cross-sectoral, multi stakeholder collaboration is vital for increasing resilience, many open questions remain.

While we have described the preconditions and procedures in several municipalities, the actual influence factors need to be discovered. Especially, we would like to understand whether some approaches of increasing resilience through collaboration outperform others. As already laid out in Sect. 3, the interrelationship of IT, communication, collaboration and resilience is complex and highly dynamic. The framework we propose helps to break down the complexity, but a further divide-and-conquer approach towards grasping the mechanisms in municipalities remains a future research topic.

Additionally, the dynamics are hardly captured. Further research needs to scrutinize time-dependence as well as typical paths towards increased resilience. We propose case study-based research to be the most appropriate for the immediate task, while quantitative work should follow and lead to verified models.

A better understanding of processes and ramifications of strategic decisions should then lead into two directions: a better theoretic understanding as well as more practical advice. These targets seem contradicting, but they merely serve two different purposes. The literature on urban and social resilience is steadily growing, yet theory building not always follows step immediately. Therefore, this gap should be closed. At the same time, municipalities seek for advice and consultancy to become more resilient – ideally avoiding unnecessary costs and time-consuming evaluation processes. If a sound theoretic understanding is gained, it should be concretized to practical advice.

4.3   Limitations

Our work poses several limitations. There are boundaries both concerning the method we employed when writing this paper and concerning the topic as a whole.

Our research paves the way towards a first framework for cross-sector collaboration in municipalities with an explicit focus on resilience. Although cross-sector collaboration in general is widely addressed in the literature, the novelty of the topic dictates that we cannot present well-established, empirically verified work at this time. Basing our work on the elaborated work with seven cities and relying on explorative, qualitative research methods mandates that we can propose very relevant and also well-argued findings; however, empirical significance remains is a topic for future activities. While we were in contact with a wide variety of different stakeholders and had access to personnel in different roles in the cities, we have not necessarily captured all possible kinds of stakeholders and roles important for cross-sectoral collaboration. Consequently, while we deem the foundation of our framework solid, it might yet have gaps when it comes to special cases, particularities, and idiosyncrasies e.g. due to national laws and customs. Although this is a typical limitation of frameworks that are not yet established, it nonetheless needs to be kept in mind. Finally, the framework is currently limited in its scope. Ideally, it should provide assessment capabilities to researchers and offer practical benefits to municipalities, e.g. in the form of accompanied *best practices*. However, again owed to the novelty of this work, to achieve both goals additional tweaking will be needed.

On a more general level, a framework – or, more precisely, a model – can only grasp the reality to some extent. It needs to be kept in mind that the framework can aid a better understanding of cross-sector collaboration and that it can facilitate discussions in this regard. It cannot be employed as a tool for direct improvements, let alone it can be expected to derive producible actions directly from the model.

Some of these limitations can be tackled, while other for the time being are fundamental to the kind of work we conducted. While the limitations do not impede the value of our work, we will try to overcome them in our future research.

4.4 Future Work

Our future work is a consequence both from the open questions that we raised and from the limitations that we named.

Following the call for more theoretical contributions while providing practitioners with advice, we will work into two directions. The above proposed framework will be discussed both with academics and with practitioners, aiming at designing a revised version and learning about the merits it already poses. At the same time, we will intensify the discussion with our city partners, aiming at providing a kind to guideline to the framework.

Considering more specific activities, we intend to focus on citizen involvement. The engagement of citizens has been acknowledged to be essential by our city partners. While this topic not directly falls under the umbrella of cross-sectoral collaboration, it has many interdependencies. Think for example of a disaster in which the population is in dire need of timely, reliable information. It cannot be straightforward answered whether this information should be provided by the municipality, the media, the police, a national authority and so on, or probably only be one of them. Collaboration, thus, plays an important role. We will seek to investigate citizen involvement in the light of communication and collaboration for resilience.

5 Conclusion

In this paper we have presented insights from a large research project on urban resilience. We have introduced the Resilience Information Portal as a main tool developed within the project and reported insights from its implementation with partner cities. Based on this, we have proposed an extension of a framework to better understand cross-sector collaboration in the context of urban resilience.

During the pilot implementation and especially during the stakeholder training workshops, it became evident that most cities are already working on improving stakeholder engagement, revising the IT systems currently in use and thinking about potential, possible ways to integrate the Resilience Information Portal's features and functionalities. In order to better integrate the Resilience Information Portal into existing communication platforms, municipal staff engaged in strategic management and the IT department of each municipality are encouraged to review the city's communication infrastructure. Although cross-sector collaboration is not the solo solution [23] to tackle all challenges that cities are facing, we believe that it can have much impact on the resilience of municipalities and that the alignment of municipal strategies will be very beneficial in this regard.

As illustrated in the prior section, our work will now continue. We seek both theoretic and practice contributions, thereby intensifying the work with the cities and extending the frameworks and theories we can provide.

Acknowledgements. This work was supported by the European Union's Horizon 2020 Research and Innovation program [grant agreement number 653569]. We would like to thank the cities as well as all interview partners for their cooperation, in particular for their openness during the work with them.

References

1. NASA: Global climate change: Vital signs of the planet (2017). https://climate. nasa.gov/effects/ https://climate.nasa.gov/effects/
2. Guha-Sapir, D., Below, R., Hoyois, P.: EM-DAT: The CRED/OFDA International Disaster Database. Université Catholique de Louvain, Brussels (2015)
3. McKinsey & Company: How to make a city great. In: McKinsey Cities Special Initiative. McKinsey & Company (2013)
4. Moss Kanter, R., Litow, S.: Informed and interconnected: A manifesto for smarter cities. Harvard Business School (2009). General Management Unit Working Paper 09-141
5. The National Bureau of Economic Research: Globalization and poverty (2017). http://www.nber.org/digest/mar07/w12347.html
6. Klein, B., Koenig, R., Schmitt, G.: Managing urban resilience. Informatik-Spektrum **40**(1), 35–45 (2017)
7. Simo, G., Bies, A.L.: The role of nonprofits in disaster response: an expanded model of cross-sector collaboration. Public Admin. Rev. **67**, 125–142 (2007)
8. Hernantes, J., Labaka, L., Gimenez, R., Maraña, P.: Revised Resilience Maturity Model. Smart Mature Resilience (SMR) project (H2020-DRS-2014 653569) Public Deliverable., November 2016
9. Serrano, N., Hernantes, J., Majchrzak, T.A., Sakurai, M.: Resilience information portal. In: Comes, T., Bénaben, F. (eds.) 14th Proceedings of International Conference on Information Systems for Crisis Response and Management (ISCRAM). ISCRAM Association (2017)
10. Grimes, C., Sakurai, M., Latinos, V., Majchrzak, T.A.: Co-creating communication approaches for resilient cities in Europe: the case of the EU project SMR. In: Comes, T., Bénaben, F. (eds.) 14th Proceeding of International Conference on Information Systems for Crisis Response and Management (ISCRAM). ISCRAM Association (2017)
11. Charmaz, K.: Constructing Grounded Theory, 2nd edn. Sage, Los Angeles (2014)
12. Grbich, C.: Qualitative Data Analysis: An Introduction. Sage, London (2007)
13. Holton, J.A.: The coding process and its challenges. In: Bryant, A., Charmaz, K. (eds.) The Sage Handbook of Grounded Theory, pp. 265–289. Sage (2007)
14. Majchrzak, T.A., Sakurai, M.: Design Principles for the Use of Social Networking Services to Promote Transdisciplinary Collaboration. Smart Mature Resilience (SMR) project (H2020-DRS-2014 653569), May 2016. Public Deliverable
15. Dyck, S., Majchrzak, T.A.: Identifying common characteristics in fundamental, integrated, and agile software development methodologies. In: Proceeding 45th Hawaii International Conference on Systems Science (HICSS-45). 5299–5308. IEEE CS (2012)
16. Aldrich, D.P., Meyer, M.A.: Social capital and community resilience. Am. Behav. Sci. **59**(2), 254–269 (2015)
17. Nahapiet, J., Ghoshal, S.: Social capital, intellectual capital, and the organizational advantage. Acad. Manag. Rev. **23**(2), 242–266 (1998)

18. Nakagawa, Y., Shaw, R.: Social capital: a missing link to disaster recovery. Int. J. Mass Emergencies Disasters **22**(1), 5–34 (2004)
19. Latinos, V., Peleikis, J.: Peer Review Meeting 1. Smart Mature Resilience (SMR) project (H2020-DRS-2014 653569), October 2016. Public Deliverable
20. Sakurai, M., Watson, R.T.: Securing communication channels in severe disaster situations - lessons from a Japanese Earthquake. In: Proceeding of Information Systems for Crisis Response and Management (ISCRAM), Kristiansand, Norway (2015)
21. Sakurai, M., Watson, R.T., Abraham, C., Kokuryo, J.: Sustaining life during the early stages of disaster relief with a frugal information system: learning from the Great East Japan Earthquake. IEEE Commun. Mag. **52**(1), 176–185 (2014)
22. Grimes, C., Latinos, V.: Report of the Review Workshop. 1. Smart Mature Resilience (SMR) project (H2020-DRS-2014 653569), October 2016. Public Deliverable
23. Bryson, J.M., Crosby, B.C., Stone, M.M.: The design and implementation of cross-sector collaborations: propositions from the literature. Public Adm. Rev. **66**(s1), 44–55 (2006)

Public Expectations of Disaster Information Provided by Critical Infrastructure Operators: Lessons Learned from Barreiro, Portugal

Laura Petersen[1(✉)], Laure Fallou[1], Paul Reilly[2], and Elisa Serafinelli[2]

[1] European-Mediterranean Seismological Centre (EMSC), c/o CEA, Bât. BARD, Centre DAM, Ile de France, 91297 Arpajon, France
petersen@emsc-csem.org
[2] University of Sheffield, 211 Portobello Road, Sheffield S1 4DP, UK
{p.j.reilly,e.serafinelli}@sheffield.ac.uk

Abstract. Previous research into the role of social media in crisis communication has tended to focus on how sites such as Twitter are used by emergency managers and the public rather than other key stakeholders, such as critical infrastructure (CI) operators. This paper sets out to address this gap by examining Barreiro residents' expectations of disaster communication from CI operators through the use of an online questionnaire and comparing the results to the current practices of the Barreiro Municipal Water Network, which were examined via an in person interview. The findings suggest that the public expect CI operators to communicate via traditional and social media and that the Barreiro Municipal Water Network should expand their current practices to include digital media.

Keywords: Social media · Traditional media · Crisis communication · Critical infrastructure operators · Public expectations

1 Introduction

Effective crisis communication can be defined as "the provision of effective and efficient messages to relevant audiences during the course of a crisis process" [1]. Social media has been identified as an increasingly important source of information during crisis situations [2]. Previous research in this area has tended to focus on how emergency response personnel or the public use social media during such incidents [3–7], overlooking other key stakeholders such as critical infrastructure (CI) operators. As such, there remains relatively little empirical research exploring public expectations of information provided by CI operators during crisis situations. The EU Horizon 2020 project IMPROVER (Improved risk evaluation and implementation of resilience concepts to critical infrastructure), makes use of Living Labs, or clustered regions of different types of infrastructure which provide specific services to a city or region. One such Living Lab is the Barreiro Municipal Water Network. This paper then addresses these under-researched issues by presenting a brief literature review on public expectations of disaster related information shared via social media. It then describes the

© Springer International Publishing AG 2017
I.M. Dokas et al. (Eds.): ISCRAM-med 2017, LNBIP 301, pp. 193–203, 2017.
DOI: 10.1007/978-3-319-67633-3_16

Barreiro case study. After, the methodology of the online questionnaire and interview-based study of the Barreiro Living Lab are described. This is followed by a presentation of the questionnaire and interview results, accompanied by a comprehensive discussion on the subject.

2 General Expectations of Social Media Use in Crisis

The public expect to be able to search for and find real-time information relating to disasters from both traditional and social media sources and research suggests that people use a combination of these sources to find information during disasters [6–9]. When it comes to social media platforms such as Facebook and Twitter, studies show that the public also expect responses from emergency services to their questions and comments [3, 6]. As previously stated, less is known about public expectations of social media use by CI operators. Self-evidently, citizens appear to expect updates from CI operators in regards to service restoration and some operators are already using social media to meet this need [2].

3 Background on Barreiro Case Study

According to the 2011 Census, Barreiro's municipality has a population of 78,764 people with an area of 36.41 km^2 [10]. The municipality is integrated in the district of Setúbal, which belongs to the Lisbon Metropolitan Area, and is located on the south bank of the Tagus River estuary. It is located about 40 km from Lisbon and is connected to Lisbon by two bridges as well as ferries [11]. The Barreiro Municipal Water Network delivers potable water to the municipality of Barreiro and serves 42,400 customers. It has an annual water flow of 6,200,000 m cubed [12]. The drinking water comes exclusively from underground aquifers [12]. Several hazards may influence the water network in Barreiro including earthquakes, droughts and heatwaves. A historical example can be found in 1969, when Barreiro's water network endured moderate damage due to a 6.8 Magnitude earthquake event which led to the unavailability of potable water for 24 h [11]. More recently, in 2012, rice and cereal agriculture in Setúbal were affected by a water shortage [11].

4 Methodology

4.1 Research Questions

Specifically, two Research Questions emerged from the literature reviewed above:

1. What do Barreiro residents expect of CI operators in regards to information provision during crisis situations?
2. How do these declared expectations compare to the current communication efforts of the Barreiro Municipal Water Network?

An online questionnaire and interview-based study was designed to investigate these questions. Ethics approval was sought and obtained from the respective authorities prior to data being collected.

4.2   Questionnaire

The target population for the questionnaire was residents of Barreiro, Portugal aged 18 years and over. In order to maximise the response rate, the questionnaire was translated into Portuguese and also made available in English. Convenience sampling was used for the questionnaire. It was structured as follows: First, a brief description of the project was provided and participants were informed of their right to withdraw from the project at any time, as well as how all data would be handled during the project. For the purposes of this questionnaire, respondents were presented with the following definition of a disaster: "an event which has catastrophic consequences and significantly affects the quality, quantity, or availability of the service provided by the infrastructure." Second, a Likert scale was used to measure participants' expectations (going from strongly agree, agree, unsure/neutral, disagree to strongly disagree). Participants were asked two questions regarding information provision. The first asked, "During and immediately after a disaster, I expect critical infrastructure operators to provide me with information…" and presented four scenarios: via calling their telephone number, on their website, on their social media site and through traditional media e.g. interviews with television networks or the radio, press releases. The second asked, "During and immediately after a disaster, I expect critical infrastructure operators to respond to my questions and comments on their social media sites e.g. Twitter." The questionnaire also asked about the participants' demographics and social media habits. Data from the questionnaire was collected between 28 March 2016 and 30 April 2016. The questionnaires were translated back into English at the data entry stage. The questionnaire was hosted on Google Forms and disseminated through the IMPROVER consortium partners' contacts and through Living Lab contacts, especially via the Barreiro Municipality Facebook page.

Sample characteristics

A total of 133 participants from Barreiro, Portugal completed the online questionnaire. Due to the dissemination method, this self-selected sample was not broadly representative of the Barreiro municipality. Sample characteristics showed that 57% of the respondents were men and 40% were women. The 2011 Census revealed a predominance of the number of women compared to men (53% against 47%) among the resident population in the Barreiro municipality [10]. Most respondents were highly educated with 69% of them having a university degree or higher, whereas only 15% of the Barreiro population has a university level education [10]. 93% of respondents are of Portuguese nationality. Both young and old people appeared to be underrepresented in the study. Respondents aged 18–24 accounted for only 4% of the total sample (for comparison, Barreiro 15–24 year olds make up 10%), with 12% identifying themselves as aged 55 years and older (for Barreiro, 22% of the population is 65 years or older)

[10]. However, the questionnaire sample is more representative of the Barreiro Face-book[1] page active users[2] when it comes to age (Fig. 1). This shows the impact of the dissemination method on our sample, keeping in mind that some respondents found out about the questionnaire via other methods than Facebook (direct ask via email, for example).

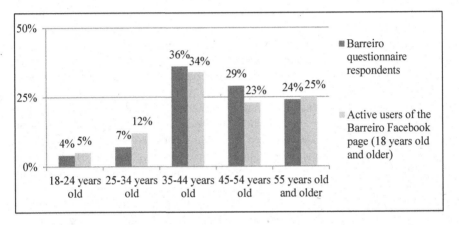

Fig. 1. Age of Barreiro questionnaire respondents compared to the age of the active users of the Barreiro Municipality's Facebook page.

Lastly, 86% of respondents have an account with a social media site such as Facebook or Twitter. For comparison, in Portugal, 70% of households have Internet access, and 69% of individuals have used Internet in the last three months [13]. When asked an open-ended question to list up to three social media sites the respondent used most, 62% listed Facebook, with 10% or less choosing other popular platforms (Fig. 2).

4.3   Interview

The findings were then compared to the current practices of the Barreiro Municipal Water Network. An in-person interview took place with actors from the Living Lab. These included an actor from the wastewater and water supply management, an environmental engineer for water quality, and the manager of the new technologies and Water Security Plan. Data from the interview was collected on 18-19 February 2016.

---

[1] The data about Barreiro Facebook was provided by Elisabete Carreira, Advisor to the Board of Directors at INOV, Portugal on 13/05/2016.

[2] Facebook defines an active user as a Facebook user who liked, commented or shared the page publication or interacted with the page in the last 28 days.

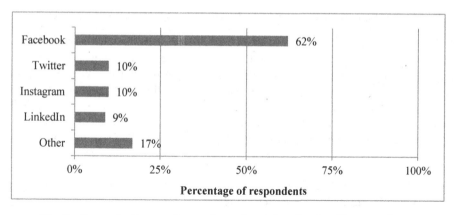

Fig. 2. Respondent's most frequently used social media sites (list up to three).

5 Questionnaire Results

5.1 Expectations for Information to Be Provided on Social Media

When asked if respondents expect CI operators to provide disaster related information on social media, 69% strongly agreed or agreed (Fig. 3). Less than a fifth of respondents (19%) were unsure or neutral in regards to the use of social media by CI operators to push disaster related information to the public, with another 12% having disagreed or strongly disagreed. Social media users had higher expectations than non-users when it comes to CI operators using this medium for crisis communication, with 74% having agreed or strongly agreed. Social media non-users were mostly (50%) unsure or neutral about this.

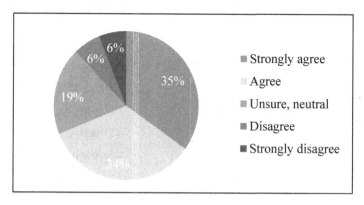

Fig. 3. Expectations that crisis information should be provided via social media

5.2   Expectations for Information to Be Provided via Other Channels

CI operators were expected to use traditional broadcast media such as newspapers, radio or television to communicate with members of the public during such incidents, with 97% of respondents having strongly agreed or agreed with this statement (Fig. 4). No respondents disagreed or strongly disagreed with this statement, and only 3% declared that they were unsure or neutral. The majority of respondents also had high expectations in relation to the availability of crisis information on the website of operators, with 79% having agreed or strongly agreed (Fig. 5). Only 14% of respondents were unsure or neutral. Respondents also expected operators to provide information via them calling a telephone number, with 77% having agreed or strongly agreed. 17% of respondents were unsure or neutral in regards to CI operators having a telephone hotline to make disaster related information available to the public (Fig. 6).

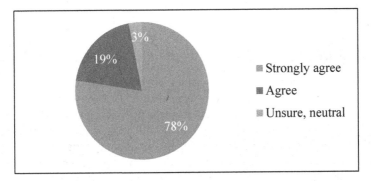

Fig. 4. Expectations that crisis information should be provided via traditional media

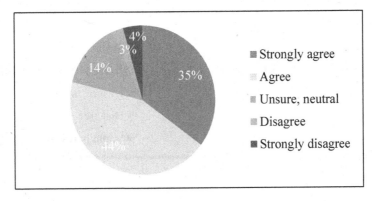

Fig. 5. Expectations that crisis information should be provided via websites

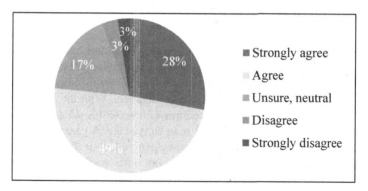

Fig. 6. Expectations that crisis information should be provided via calling a telephone number

5.3   Expectations for Two Way Communication on Social Media

When asked if they expect CI operators to respond to questions and comments sent by members of the public to their social media accounts, 63% of respondents agreed or strongly agreed (Fig. 7). There was high uncertainty/neutrality (18%) in regards to responding to queries, and a large portion of respondents selected disagree and strongly disagree (14% and 5%, respectively). Social media users have a higher expectation than non-users in regards to responding on social media, 68% agreed or strongly agreed. For social media non-users, the majority (39%) were unsure or neutral, and these respondents appeared to be more divided on the issue with 34% having agreed or strongly agreed compared to 28% having disagreed or strongly disagreed. Lastly, 58% of respondents who expected CI operators to communicate via social media also expected them to respond to questions or comments.

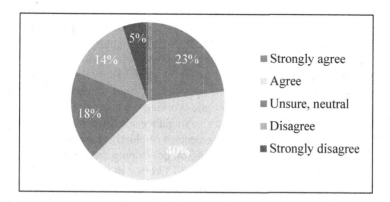

Fig. 7. Respondents' expectations for two way communication on social media during and after a disaster

6 Interview Results

The Barreiro Municipal Water Network considers communication with their stakeholders as not very efficient, stating that they do not communicate with the general public enough. Currently the operators do not have their own communication strategy in place to communicate with the public during disasters. With the implementation of Water Safety Plan they are creating procedures to reverse this situation and develop a real communication strategy, but it is still at an early stage. A general communications plan does exist within the municipal emergency plan, which includes using traditional media. The Barreiro Municipal Water Network also participates in semi-annual meetings within the framework of the municipal observatory for water issues. The Water Safety Plan will mandate that they have stakeholder meetings with for example regulators, municipal services, civil protection, local charities, the industrial park companies, local hospitals and the general public. These face-to-face stakeholder meetings are seen by the operators to be the most effective means of communicating with the public. However, the most used means of communication is the telephone. Indeed, they provide a call center for damages. The operators also believe that most people in the city would use the telephone to contact the emergency response authorities if there was a disaster. The operators do not have their own website or social media accounts.

However, it appears that most of the communication with the public comes from the municipality in general, and not from the infrastructure operators themselves. Indeed, the operators participate in neighbor meetings that are organized for the entire municipality where water is not the only issue being discussed. The municipality also has a media department that handles communication with the public and they work closely with the operators. The municipality provides a website, an FAQ page, and social media sites. Despite this, the operators said that they feel that "communication tends to fail during incidents."

The operators only inform the population of the interruptions that are scheduled and that are more time consuming and involve a higher level of complexity in terms of implementation. These interventions are usually done at night to cause minimal impact. However, and although not always, but in many cases, when it comes to service disruptions, it has been the citizens who report by calling their telephone line. "People are searching for information," said one interviewee, "and we recognize the need to become more proactive." As such, they are working on a smartphone application to provide the public with information and also to gather information from the public. As the municipality of Barreiro has a high percentage of elderly residents that do not have access to the internet, they are also focusing on developing automatic communication via SMS to reach those residents. They would like for in the future to have a clear strategy in terms of communication, internal and external, that guaranteed the population that their basic needs will be met, in order to strengthen public confidence in the authorities with emergency situations and to guarantee a collaborative platform among all stakeholders, including the general public. The operators also mentioned that Portuguese people are known for their resilience and adaptability, which means that current communication expectations are probably not as high as they could be.

7 Discussion

Results indicate that members of the public expect CI operators to provide disaster related information via both traditional and social media. Overall, expectations were high for all four channels studied. There was no disagreement among respondents when it came to expecting disaster related information relevant to CI to be provided by CI operators through traditional means such as radio or television. This further reinforces the idea that even to the public, social media is meant to compliment more traditional crisis communication methods and not replace them. Social media should then be used as another channel for information dissemination, as 69% of respondents expected it to be used by CI operators. High expectations for information to be available on the Internet via social media and websites demonstrate high expectations for operators to be proactive in pushing information to citizens. The majority of respondents expected operators to respond to queries, confirming the importance of two-way communication. However, the high amount of disagreement with this statement furthers the idea that pushing information to citizens should be more of a priority for operators than responding to queries. People tend to use the media platforms that they are already familiar with during crisis [14, 15], and as such it seems natural that respondents who use social media have higher expectations for social media use by CI operators than social media non-users. However it is interesting to note that despite a telephone number being the only current, and thus familiar, way of contacting Barreiro Municipal Water Network, expectations for information to be available on a website were higher than for there to be a telephone hotline. This further substantiates the importance of understanding the technological culture of the local population when putting into place crisis communication plans [16].

The current efforts of the operators at Barreiro Municipal Water Network to communicate with the public appear to meet certain public expectations. They currently use traditional media and have a telephone hotline available for the public to use, both of which were highly expected from respondents. However, as they do not have social media accounts or a website, they are not yet meeting this expectation. As the IMPROVER questionnaire did not establish what the participants understood as crisis-related information, this expectation may be an uninformed one. The fact that expectations for information to be available on a website were higher than those for a telephone hotline clearly indicates that this is an area where the operators can improve the effectiveness of their crisis communication. As they already work alongside the municipality more generally when it comes to public communication, it may be interesting to see if the municipality's social media sites and websites would be considered by the public as an appropriate source to find CI related disaster information. More research should examine if the CI operators need to use social media, or if they just need to make their information available via social media via other official sources. Further research should look into this. While the operators find that face-to-face stakeholder meetings are the most effective means of communicating with the public, this option was not presented to respondents as it is impractical to have a stakeholder meeting in the immediate aftermath of a damaging disaster. Despite the fact that they do not currently use Internet to communicate with the public, their acknowledgement of the need to be more proactive in informing the public in times of crisis demonstrates an awareness of these issues. Their ideas to develop smart phone applications or SMS alerts are innovative, but as these two channels were not presented as options to respondents, further work should study public expectations for these two channels. In conclusion, it is suggested then that the operators expand their crisis communication strategy to include social media and websites.

7.1   Limitations

The method we followed is not free from limitations. As discussed earlier, a self-selecting sample group was involved in the questionnaire study, which did not adequately represent the demographics of Barreiro. It should also be noted that the use of the website to distribute the questionnaire was likely to have skewed the sample in favor of those who used the Internet and social media on a regular basis.

8   Conclusions

Our findings suggest that CI operators should continue to use traditional media during crisis situations, and that this should be supplemented through the provision of disaster related information via websites, telephone hotlines and social media platforms maintained by CI operators. The Barreiro Municipal Water Network should expand their crisis communication efforts to include Internet communication via social media and websites in order to meet public expectations. However, it should be acknowledged that this was a self-selecting sample that was not representative of the demographics in the population studied, and as it was an online questionnaire it most likely attracted people who generally use the Internet and social media. Further work is needed to explore the perspectives of citizens who are unable or unwilling to use digital media. A future questionnaire is being developed within the IMPROVER project that will be disseminated by telephone in Barreiro in order to try to reach such citizens and it will also address expectations of smart phone applications and SMS as communication channels for CI operators. This work should also examine the differences between expectations of CI operators as information providers and the expectation of CI related information to be available. Other future work will consider the implications of these findings for the development of the Barreiro Municipal Water Network communication strategy and Water Safety Plan. Lastly, the IMPROVER project is also currently working on the development of a communication strategy for CI operators to deploy during each stage of a crisis that will encompass both digital and traditional media platforms.

Acknowldegements. The IMPROVER project has received funding from the European Union's Horizon 2020 research and innovation programme under grant agreement No 653390.

References

1. Freberg, K., Saling, K., Vidoloff, K.G., Eosco, G.: Using value modelling to evaluate social media messages: the case of Hurricane Irene. Public Relat. Rev. **39**, 185–192 (2013)
2. Petersen, L., Fallou, L., Reilly, P., Serafinelli, E.: Public expectations of social media use by critical infrastructure operators in crisis communication. In: 14th ISCRAM International Conference on Information Systems for Crisis Response and Management Conference on Agility is Coming, IMT Mines Albi-Carmaux (École Mines-Télécom), Albi, France (2017)
3. Reuter, C., Spielhofer, T.: Towards social resilience: A quantitative and qualitative survey on citizens' perception of social media in emergencies in Europe. Technol. Forecast. Soc. Chang. (2016). doi:10.1016/j.techfore.2016.07.038
4. Reuter, C., Ludwig, T., Kaufold, M., Spielhofer, T.: Emergency services' attitudes towards social media: a quantitative and qualitative survey across Europe. Int. J. Hum Comput Stud. (2016). doi:10.1016/j.ijhcs.2016.03.005i
5. Bruns, A., Burgess, J., Crawford, K., Shaw, F.: # qldfloods and @ QPSMedia: Crisis Communication on Twitter in the 2011 South East Queensland Floods. ARC Centre of Excellence for Creative Industries and Innovation, Brisbane (2012)
6. The American Red Cross: Web Users Increasingly Rely on Social Media to Seek Help in a Disaster. Press Release (2009)
7. Lindsay, B.R.: Social Media and Disasters: Current Uses, Future Options, and Policy Considerations. CRS Report for Congress (2011). https://www.fas.org/sgp/crs/homesec/R41987.pdf
8. Burger, J., Gochfeld, M., Jeitner, C., Pittfield, T., Donio, M.: Trusted information sources used during and after Superstorm Sandy: TV and radio were used more often than social media. J. Toxicol. Environ. Health **76**(20), 1138–1150 (2013)
9. Mitomo, H.; Otsuka, T.; Jeon, S.Y.; Cheng W.: The role of ICT and mass media in post-disaster restoration and recovery progress: a case of the Great East Japan Earthquake. In: 24th European Regional Conference of the International Telecommunication Society, Florence, Italy, 20–23 October 2013 (2013)
10. Statistics Portugal (INE - Instituto Nacional de Estatística). https://www.ine.pt/xportal/xmain?xpgid=ine_main&xpid=INE
11. Ioannou, I., Aspinall, W., Bouffier, C., Carreira, E., Honfi, D., Lange, D., Melkunaite, L., Kristine Reitan, N., Rossetto, T., Storesund, K., Teixeira, R.: IMPROVER Deliverable 2.1 Methodology for Identifying Hazard Scenarios to Assess the Resilience of Critical Infrastructure (2016)
12. Barreiro Municipality: Gota a gota: Agua no barreiro: um tesouro Escondido. A água e o saneamento no concelho do Barreiro, 1 March 2009
13. Eurostat: digital economy and society statistics - households and individuals (2015). http://ec.europa.eu/eurostat/statistics-explained/index.php/Information_society_statistics_-_households_and_individuals. Accessed 20 April 2016
14. Fire Services Commissioner Victoria: Information Interoperability Blueprint (2013). http://fire-com-live-wp.s3.amazonaws.com/wp-content/uploads/Information-Interoperability-Blueprint.pdf
15. Steelman, T.A., McCaffrey, S.M., Velez, A.L.K., Briefel, J.A.: What information do people use, trust, and find useful during a disaster? Evidence from five large wildfires. Nat. Hazards **76**(1), 615–634 (2014). doi:10.1007/s11069-014-1512-x
16. Fallou, F.: Culture and Risk management in Man-made and Natural Disasters. EU H2020 Project CARISMAND (2017)

Secure and Privacy-Preserving Information Sharing for Enhanced Resilience

Jaziar Radianti[1](✉) and Terje Gjøsæter[2]

[1] Department of ICT, Centre for Integrated Emergency Management UiA,
Jon Lilletunsvei 9, 4879 Grimstad, Norway
jaziar.radianti@uia.no
[2] Department of Computer Science, HIOA,
Box 4 St. Olavs plass, 0130 Oslo, Norway
terje.gjosater@hioa.no

Abstract. City resilience is a pressing issue worldwide since the majority of the population resides in urban areas. When disaster strikes, the consequences will be more severe in the cities. To achieve resilience, different entities and the public should share information during a disaster. ICT-based Community engagement is used for strengthening resilience. We suggest easy-to-use metrics for assessing the security and privacy of information sharing platforms for resilience, and apply them to selected platforms. Most of them are reasonably well-protected, but with less than private default settings. We discuss the importance of security and privacy for different important categories of users of such systems, to better understand how these aspects affect the willingness to share information. Security and privacy is of particular importance for whistle-blowers that may carry urgent information, while volunteers and active helpers are less affected by the level of security and privacy.

Keywords: Resilience · Security · Privacy · Resilience tool · Information sharing

1 Introduction

The UNDESA projection shows that the proportion of the world's population living in urban areas will increase from 54% (2014) to 66% by 2050 [1]. Thus, it is evident why city resilience has been emphasized globally due to cities becoming more vulnerable as more people will be affected when unexpected events occur. Information sharing is an important way to enhance resilience in a disaster [2, 3], with the help of information and communication technologies (ICT) tools that are becoming acceptable for facilitating crisis communication [4]. Interpreting further the spirit of the Hyogo Framework Action [5], the overall capability to cope with hazards is not solely authority responsibility, but is a combination of the self-organizing capability of the individuals, communities, public and private organizations in affected areas. Hence, the role of ICT tools to enable the society in general to adapt and recover from hazards and stresses is evident [6], especially to ensure that the right information flows smoothly to the intended audience.

I.M. Dokas et al. (Eds.): ISCRAM-med 2017, LNBIP 301, pp. 204–219, 2017.
DOI: 10.1007/978-3-319-67633-3_17

What is resilience? UNISDR [7] defines resilience as "the capacity of a system, community or society potentially exposed to hazards to adapt, by resisting or changing in order to reach and maintain an acceptable level of functioning and structure". We adopt this definition, and suggest that resilience should include the ability of individuals and communities to absorb and prepare to make use of the different crisis management communication technologies, so that they can engage and share information better with each other and with public authorities. The definition should also include the capacity for learning practical security and privacy knowledge for better use of ICT-based engagement tools.

The recent trend of ubiquitous computing allows people to share information by using day-to-day technologies surrounding them. The presence of social media in combination with the powerful trend of crowdsourcing for obtaining data shared by citizens [8, 9], has to some extent been accepted as a part of crisis communication, apart from the data quality weaknesses that may arise from social media information [10]. Sharing information using various means arguably improves resilience as individuals can contribute information faster to the authorities, as well as to the circle of family and friends they care about and improve the way responders manage the crisis [6, 11].

However, the crucial questions that should be addressed are: how thoroughly are the privacy and security concerns considered in line with the encouragement of the information sharing among different components of a resilient society? Does information sharing increase the resilience, or could it in some situations weaken the resilience, when a focus on security and privacy emerge?

Information security concerns protecting information in different contexts: its confidentiality, integrity and availability [12]. Security of information is essential to some organizations and actors before they are willing to share it [13]. For private citizens, trust concerning privacy protection is crucial, covering personal sensitive information (PSI) and personally identifiable information (PII, information that can identify them) [14]. Part of the resilience is that our information is verifiably unmodified, confidential, available when needed, and accessible to authorized personnel only. These may in some cases seem to be sacrificed or at least prioritised down in a disaster situation, but should not be, and indeed does not need to be. On the contrary, the negligence of security and privacy could in fact harm the information sharing by making some actors reluctant to share potentially important information.

This paper discusses a selection of techniques, technologies and tools that are used for information sharing in some countries, or general tools that are used worldwide such as social media. We discuss how individuals and communities can be more resilient in a crisis, in terms of the way they share information, by addressing security and privacy concerns.

This paper is divided into 6 sections. In the next section we describe our scope, research questions and methodology. Section 3 describes the relevant literature for our case. In Sect. 4, we report the result and analysis from our study. Discussion and lessons learnt from our research and implications for disaster resilience is presented in Sect. 5. Section 6 is a summary of the main findings of this study and its limitations.

2 Scope and Methodology

To clarify the limits and focus areas of our research, we delineate the scope of this article. First, we will not discuss privacy and security in terms of algorithms which are very common in computer science literature [15–17]. We also did not include the privacy issues caused by providers of engagement platforms. Instead, we suggest metrics that are possible to use for people with limited computer security background to assess security and quality of engagement tools or platforms. In other words, we use testable, easy-to-use metrics. Finally, in the outcome of this study, we will not propose metrics to measure the influence of privacy and security on the disaster resilience. We will instead discuss the properties of three different user groups regarding the willingness to share information in a disaster. From this point of view, we will then discuss qualitatively how these properties affect resilience with examples or scenarios. The central research questions (RQs) in this paper are as follow:

RQ1: What is a good pragmatic approach to evaluate security and privacy of tools and platforms for citizen engagement in disasters?

RQ2: What aspects of information sharing for supporting disaster resilience are not well covered in current state of the art?

RQ3: In what way can security and privacy concerns strengthen or weaken the disaster resilience?

There are several practical examples behind our earlier questions and arguments. Implementation of encryption, for example, a common method to protect information, is resource consuming. If the security is very strong and complex, it can make implementation and execution of the system hard, and usage more complicated. The end result could be that good intentions lead to making information less available rather than more. To support our research, we will investigate several cases and examples, as well as carry out a thorough analysis on these cases to show the relevance of our research questions. We will also support this by looking at the current information sharing tools and providing scenarios where security and privacy can be highly important, but currently often overlooked.

The contributions of this article are fourfold. First, we propose evaluation criteria for security and privacy of information sharing tools that are commonly used, or designed for community engagement and information sharing purpose in a disaster situation. Second, we perform an evaluation of a relevant selection of tools according to our evaluation criteria and for some selected cases. The method is non-intrusive, based on published information, documentation, and policies. Third, we suggest practical recommendations for stakeholders wanting to implement a new engagement tool. Fourth, we define groups of users as a starting point to enable us to discuss properties of groups that can contribute to strengthening the city resilience, and to discuss how to build synergy and minimize trade-offs between security, privacy and resilience.

We use a three-stage procedure, i.e. investigating different metrics for evaluating security and privacy, reviewing a selection of information sharing tools to test our proposed evaluation metrics, and finally examining different typical user groups and their needs for security and privacy to be willing to share information. We use the results as a basis for coming up with a set of recommendations. Our methodology is as follows:

Stage 1-Metrics. We investigate different methods or metrics for evaluating the *security* and *privacy* of information sharing tools, and then select methods that are non-obtrusive, allow us to observe without being an insider. There are many criterias that could be used to evaluate the *security* of the systems in question, among others:

- *Security by design/built-in security*: This is a common criteria for evaluation of security [18]. The system should be designed and built with security as a fundamental requirement from the start, not as an afterthought. However, this criterion is hard to judge in our case because we are doing black-box evaluation. In other words, security by design is only possible to evaluate with some degree of certainty for insiders, and is not easy to judge without knowing the technical details of the tool. Therefore, this metric is not included in our evaluation.
- *Aspects of Security*: Security has different aspects that can be discussed separately [12]: confidentiality, integrity and non-repudiation, and availability. It is important that these aspects are covered by the selected metrics. Authentication is also essential to ensure that the user is authorized to access information.
- *Testable security criteria*: Based on the listed security aspects as shown in Table 1, we select the following set of criteria for security evaluation since they are immediately observable and testable as a user on the running platform.

Table 1. Principles of security

Security aspect	Tested
Confidentiality	Secured communication (https/ssl/tls)
Integrity and nonrepudiation	Secured communication (https/ssl/tls)
	Can messages be deleted or modified
Availability	System is available at time of testing
Authentication	Password strength requirements
	2-factor authentication available

Concerning privacy, the Privacy by Design concept is essential. It is based on seven "foundational principles" [19] as seen in the Table 2. The factors that can be tested to give an objective result in our scenario are marked as such in the table:

Table 2. Principles of the privacy by design

Principles	Testable
Proactive not reactive; preventative not remedial	
Privacy as the default setting	X
Privacy embedded into design	
Full functionality – positive-sum, not zero-sum	
End-to-end security – full lifecycle protection	X
Visibility and transparency – keep it open	
Respect for user privacy – keep it user-centric	

Several of these criteria are vague and hard to give a binary score, and some require insider information. And, not all security requirements can be tested [20]. Therefore, we have made a selection of the privacy and security criteria, and concretized them into the tests that can be seen in the Table 3. The definitions and applications of these metrics are provided in Sect. 4.

Table 3. Testable metrics for our research purpose

Metrics	What to check	Privacy/security aspect
Encrypted communication	Communication over https/ssl/tls	Confidentiality, integrity
Password minimum requirements	e.g. requirements for minimum length, combination of characters	Authentication
Optional 2-factor authentication	Extra factor in addition to password, e.g. one-time code from mobile app or SMS	Authentication
System is online at time of testing	Simply testing that it is possible to use the system at time of testing	Availability
Privacy policy statement on web page or in app	Privacy policy statement visible from home web page or app start screen	Privacy
Privacy configuration available	Is it possible to modify the privacy settings for the user?	Privacy
Privacy-preserving options as default setting	If privacy configuration is available, do the default settings preserve privacy?	Privacy

Stage 2-Tools. In this stage, we selected samples of tools for information sharing, as there are many tools or platforms available for use in non-crisis and crisis situations. In each stage of the emergency management cycle (preparedness, response, recovery and mitigation), different information sharing tools may be used. Which tools are preferred to use can also vary from country to country, and from hazard to hazard. A wide range of ICT tools, and technologies has been proposed and used, ranging from Wiki platforms, Smartphone apps, social media (especially Facebook, YouTube and Twitter), and other online engagement and real-time community mapping platforms.

Ushahidi [21] or Google Crisis Response [22] are examples of platforms for community mapping. Some of these ICT-based tools support crowdsourcing. In different countries, smartphone apps for emergencies have been widely used as communication tools by the government such as FEMA App [23], Hurricane App (USA), Disaster Alerts, Emergency+, First Aid or Fire Near Me [24] (Australia). Globally, some apps have been developed to alert of earthquakes such as QuakeWatch [25], Earthquake buddy [26] or Disaster Alert [27].

For testing purposes, we have looked at different categories of information sharing models to support crisis, i.e. Wiki-based platforms, a large amount of mobile-apps, and social media, community mapping. We have only included those that are formally adopted, or recommended as crisis communication tools in a specific country, or region. We varied the geographical area of the origin of the tools, and included globally popular social media. The availability of these tools for testing, and enabling citizen engagement, were additional criteria we used when searching for them, thus we omitted the commercial ones. It is worth to mention that it is not our intention to provide an exhaustive list of engagement tools. Our goal is rather to provide exemplary cases where it is possible to use and test our proposed criteria to evaluate the security and privacy matters. The list of the tools covered in our analysis is as follows:

Wiki-based Platforms. We have selected the two platforms *Wiki for professionals* [28] and *Emergency 2.0 Wiki* [29]. *Wiki for professionals* is a product from the EU FP7 PEP (Public Empowerment Policies in Crisis Management) project that tried to engaged public to take concrete actions and share information in crisis preparedness, planning and response. Inclusion of public communication initiatives in authority communication, accessibility and inclusiveness of authority communication and making information widely available and findable, are among the strategies that are consider by the PEP project as key enablers for public empowerment.

Mobile Apps. *FEMA App*: The app can be used for sharing disaster pictures, save a custom list of the items in your family's emergency kit, as well as the places users will meet in case of an emergency, and locations of open shelters. *Emergency+* is a national app circulated by Australia's emergency services to enable people to call the right number at the right time, anywhere in Australia. The app uses a mobile phone's GPS functionality so callers can provide emergency call-takers with their location information as determined by their smart phone. Emergency+ also includes SES and Police Assistance Line numbers as options, so non-emergency calls are made to the most appropriate number.

Social Media. A study in the Public Empowerment Policies [30] project shows that Facebook, Twitter, blogs and Youtube are most preferred social media for preparedness, response and recovery. For our testing purpose, we look at Twitter, YouTube, Facebook and Google+ that are popular channels for communication, also about disasters.

Community Mapping. These tools are often used with a crowdsourcing approach to information sharing or participatory mapping. A group of digital volunteers works together in a common shared map intended for improving the knowledge about disaster information, such as location of shelters, victims, hospitals, the supply needs. Ushahidi and Google Crisis Response, will be used as examples in this article for further analysis. These metrics and tools will form a basis for answering **RQ1** and **RQ2**.

Stage 3-Use Cases. In the third stage, we examine the use cases in more detail, i.e.:

- *Whistle-blowers:* ("The dam is going to break, and the manager wants to hush it down instead of evacuating the valley!"). Whistle-blowers have a strong need for protection, or even their physical security could be endangered.
- *Social Media Users*; twitterers and other social media members writing information that is accidentally or intentionally relevant for a case but aimed at friends/family and harvested by some tool. The general social media using public expect a certain level of security and privacy in the social media platform, and they should be able to expect their privacy to be respected if their posts are harvested for emergency management use, e.g. by anonymization or aggregation.
- *Active Helpers* and Disaster Actors i.e. People entering information into a tool with the express purpose of mitigating the disaster and strengthening the resilience.

They know what they are doing, and in most cases, we only need to provide a minimum of security and privacy.

These three groups of users will be central to discuss **RQ3** - if privacy and security strengthen or weaken resilience (Sect. 5).

3 Related Works

This section targets answering the following questions: How do information sharing to increase resilience appear in the literature, and how are security and privacy discussed in the resilience context? What kinds of gaps exist when discussing community engagement, the use of technologies, security, and privacy?

Indeed, sharing information is important, but recognizing factors [2] and challenges [31] that influence the success of information sharing is even more crucial. Different studies have mentioned that motivations, approaches and channels affect successful information sharing and indeed the technology. As often discussed in the Technology Acceptance Model (TAM), acceptance of technology is actually influenced by many factors [32–34], such as usefulness, being easy to use, trust (in giving personal information, in the technology itself), subjective norms, perceived innovativeness, and many more. Likewise, acceptance of the use of technology for communication in emergencies and sharing are affected by similar factors [35]. Note that trust is, in fact, one of the main factors that influence whether or not people will share information.

At this point, the importance of security and privacy will gradually come into the resilience picture via the following sequence of cause-effects: the resilience of the city is built upon community resilience which is basically the engagement of individual citizens in a disaster. Community resilience itself is built upon the willingness of individuals and organizations to cooperate and share information through existing or planned communication channels that more and more relies on ICTs. Attitude to privacy is very personal [36], whether or not anonymity matters for them, thus security and privacy will be important. For some groups of individuals in the society, lack of security and privacy reduces enthusiasm for sharing information. The chain of weakening reverse effects will eventually divert the city's goal from achieving resilience.

Measuring security and privacy is not trivial, as a lot of metrics have been proposed [37], and many of them are not so easily used if one is not a computer security expert. Pekárek and Pötzsch [38] and Hull, Lipford [39] addresses the privacy issues in collaborative workspaces and social networks, which also can be including the consent dilemma [40]. Pekárek and Pötzsch [38] compare the Wikipedia and Facebook platforms and point out that on both platforms, it is quite simple for third parties to gain access to personal data without infringing the technical rules set out for the use of the systems. In the case of Facebook this is due to the belief on the privacy default settings are optimum, or the users have no interest in privacy settings at all. For users of Wikis, customisation is simply not foreseen by the application, and thus the general user is often allowed access to a limited set of personal information, i.e. a basic profile or the user page. In the meantime, Hull, Lipford [39] discuss further Facebook privacy issues arising from features, allowing non-friend users to see the contents shared for specific friends. The privacy design is blamed as a cause of this issue. In brief, privacy and security issues that may arise from different information sharing tools are evident, but in fact, it will also depend upon what types of users that will use the tool.

4 Results and Analysis

In Sects. 4 and 5, we answer the three questions posed in Sect. 2. First, we consider *RQ1: What is a good pragmatic approach to evaluate security and privacy of tools and platforms for citizen engagement in disasters?*

To answer this question, we have proposed a set of metrics that are testable from the user perspective. It means that an organisation of institution can quickly evaluate sharing platform tools that are available in the market (as free, open-source or commercial tools). We investigate a selection of existing solutions or platforms that are already in place so they can be tested according to the selected criteria. The definition of each metric used for evaluation is shown in Table 4.

Table 4. Security-privacy metrics and definition

Metrics	Definition
Secure communication	If the tool is accessed through a secure connection (https) or not
Password requirements	If there are requirements to the password strength used to log in and used the tools or services, e.g. minimum 6 characters, mix of letters and numbers, or have to include special characters
2-factor authentication	If the users need to provide additional authentication in addition to the password, e.g. a code sent to the mobile phone
Availability	If the service is available at the time of testing
Privacy policy	If there is a clear privacy policy statement available
Configurable privacy	If users have a freedom to decide which personal information they are willing to share into the platform
Privacy as default	If the default setting of the privacy is public or private. For example, the default setting of the Facebook profile picture is open to the world. Users who are not aware of this default setting, may accidentally share his/her picture although it was not the initial intention
Asking unneeded personal info	If the tool asks unnecessary personal info when registering for the service, such as birth date
Modify or delete after reporting	If the tool allows modification or deletion of a message after it is submitted

The evaluation results of the security and privacy of different tools using our selected metrics is presented in Table 5. The row lists the tested tools, while the columns captures the metrics used for testing. We notice that one aspect that is not well covered is how the usability is affected by the security and privacy. Is the tool more complicated to use because of the increased security and privacy? One of the metrics

Table 5. The test results of the security and privacy of the information sharing tools

TOOLS	Security				Privacy				Non-repudiation
	Secure communication	Password requirements	2-factor authentication	Available at time of testing	Privacy policy statement	Configurable privacy	Privacy as default	Asking unneeded personal info	Modify/delete after reporting
Social Media									
Twitter	Yes	Yes	Yes	Yes	Yes	Yes	No	Yes	Yes (delete)
Facebook	Yes	Yes	Yes	Yes	Yes	Yes	No	Yes	Yes (marked)
Google+	Yes	Yes	Yes	Yes	Yes	Yes	No	Yes	Yes
YouTube	Yes	Yes	Yes	Yes	Yes	Yes	No	Yes	Yes (delete)
Wiki-based Platform									
Wiki for professionals	No	No[1]	No	Yes	No[2]	No	No	No	Yes (log)
Emergency 2.0 Wiki	?[3]	?	?	Yes	Yes[4]	?	?	?	?
Mobile App									
FEMA App	?[5]	No	No	No[6]	Yes	No	Yes	No	No (moderated)
Emergency+	N/A[7]	N/A	N/A	No[8]	No	No	No	N/A	No
Community Mapping									
Ushahidi deployments[9,10]	Optional[11]	N/A	N/A	Yes[12]	Yes[13]	Yes[14]	Yes	Optional[15]	No
Google Crisis Response	Yes	Yes	Yes	No[16]	Yes[17]	No	No	No	Yes
Facebook safety check	Yes	Yes	Yes	No[18]	Yes	Yes	No	Yes	N/A
Instant Messaging									
Skype	Yes[19]	Yes	No	Yes	Yes	Yes	No	Yes	Yes (marked)

Note: 1) Allows e.g. "123" which is a very weak password.; 2) Link present, but no text.; 3) Test user activation pending; 4) Activated LinkedIn group membership; 4) Via Terms of Use; 5) No information if photo upload is secure; 6) No personal identifiable information (PII) sent; 6) Only available in USA.; 7) Only for making phone calls; 8) Only available in Australia; 9) https://beinglgbtinasia.crowdmap.com;10)Deployment in Sweden http://www.diskrimineringskartan.se; 11) Depends on deployment. No PII sent by default.; 12) Deployed when needed; 13) Depends on deployment; 14) For deployment managers, not for end-users; 15) Anonymous allowed; 16) Deployed as needed; 17) Basic warning info only. 18) Activated when/where needed; 19) Call from skype to phone is not encrypted across the phone network.

touches on this, regarding the good defaults for privacy. If one has to modify several complex settings to get the system into an acceptable state regarding privacy, as is the case in particular for social network sites, the usability obviously suffers. However, to capture a more complete picture of this issue would require a much more resource-consuming user testing, and is outside the scope of this study which is focused on simple-to-test metrics. The results in the Table 5 are used to answer *RQ2: What aspects are not well supported concerning the information sharing for supporting disaster resilience in current state of the art?*

We see that the tests to all three social media options give almost the same results. In general, they are reasonably well-protected from a security point of view, and allow the users to control their privacy - however with less than private defaults. In addition, the users may retract their messages without trace, and in the case of Facebook, edit a message after posting - with an indication that the message has been edited. Note that Twitter also has guidelines for use in crisis situations [41]. The wiki-based platforms are much weaker on security, having unencrypted communication and little or no requirements for passwords. There also seems to be little focus on privacy, and indeed the wiki concept is all about openness and information sharing.

On the other hand, all changes are logged, so information cannot be retracted undetected once posted. Note that the Emergency 2.0 wiki requires manual steps to add a new user, so we have not been able to test this as thoroughly as the other tools and platforms. The mobile apps are both limited to their respective national audiences, and our results therefore depend on what can be glanced from public documentation and descriptions. Among the community mapping platforms, Ushahidi is special in that it is not one single tool, but rather a platform to be deployed in time of need (e.g. in Nepal after the earthquakes in spring 2015), and therefore we have sampled a selection of different Ushahidi installations to capture a representative impression on which we base the results. What is particularly interesting about Ushahidi is that it allows anonymous messages, without the creation of an account, as opposed to most of the other tools and platforms. Google crisis response consists of several tools, we have chosen to evaluate the *person finder*. As no fully operational person finder was available at time of testing, we base our results on a test setup. In the same way, Facebook safety check is only made available in particular large-scale emergencies, and only for people in the affected regions, so it is not possible to test. Therefore, these results are based on information gathered from documentation and other relevant sources.

5   Discussion, Solutions and Implications for Resilience

In this section, we will answer *RQ3*: *In what way can security and privacy concerns strengthen or weaken the disaster resilience?* We analyse the willingness of different groups (whistle-blowers, social media users and active helpers) to share information during a crisis based on each group's preference on required security and privacy strength. A city is used as exemplary case in our analysis. Our discussion will focus on three points: (1) Situations that will strengthen or weaken the resilience, based on user group perspectives; (2) The predicted preferable tools of each group; and (3) The information flow model based on the tested tools linked to the predicted need for security and privacy, and user group categories that are suited for each information flow model.

Table 6 depicts the proposed framework to analyse the willingness to share information given different privacy and security strength in the engagement platforms. The rows represent the user groups. The columns capture the strength categories of security and privacy embedded in the sharing tools i.e. "No privacy/security", "Average privacy/security" and "Strong privacy/security/anonymity". *The light grey* area on the right side represents the optimal smooth information flow to the city

Table 6. Willingness to share information

Users	Tools		
	No privacy/ security	Average privacy/ security	Strong privacy/ security/ anonymity
Whistle-blower			X
Social media users		X	X
Active helpers	X	X	X?

stakeholders, when the preferable privacy of users match the provided information sharing platforms. *The sark grey colour area* in the middle, shows the information flow to the city when the security and privacy level of the tools is average. While *the black area* in the left side is a situation where only people who do not bother so much about privacy, motivated by altruistic spirit and would just help facilitating the communications. In this situation, we may lose the potential information from two other groups, i.e. *Whistle-blowers* and *social media users*.

Table 6 implies that active citizen engagement for sharing disaster related information only occur if the stakeholders can provide tools that incorporate different groups' requirement for security and privacy. The grey area in Table 6 represents the information flow from different user groups that may be weakened or blocked.

Strong privacy could also include anonymity, which will encourage *Whistle-Blowers*, since they can submit reports without risk of repercussions. If we consider our analysis in Sect. 4, the *Whistle Blower* will tend to use e.g. an Ushahidi-type platform where reporters can provide information without being identified or required to login. However, Ushahidi, of course, is very much dependent upon the preference and deployment configuration of the platform owners if they would like to encourage submission of information from *Whistle-Blowers* or only from *Active Helpers*.

Why do we care about *Whistle-Blowers* in this information sharing context? Because *Whistle-Blowers* who want their privacy to be particularly protected, could be the group that possess unique and important information that may require rapid handling and mitigation. Therefore, they have a clear reason and need to be protected as informants. Wikileaks [42] is an extreme example of framework that fits the *Whistle-Blowers*, where people feel secure to share information anonymously without fear of being identified as a reporter, apart from the controversy surrounding this case. In the disaster case, the example could be any extreme hazards such as industrial disaster hazards e.g. chemical leaks, radiation leaks to the water system or other critical infrastructure services that is vital for the city life and the citizens. In such case, the most knowledgeable person knowing the detail of the case may be reluctant to openly share the information because of many different reasons such as loss of reputation, job or even being taken to court for leaking confidential information.

The *Active Helpers* may not care about strong or weak security because the motivation is to help, share information and contribute as much as possible to mitigate the disaster impact. Thus, too much security may just hinder or slow them down to actively share information, which eventually may weaken the resilience. Thus, the X sign with a question mark in the right bottom corner in the Table 5 represents the double-edged sword issue that may arise, when the extra effort to ensure security becomes too much, while this group could in fact be the most active one.

By having a good framework for understanding the willingness to share in the different groups of users as shown in Table 6, we can then predict the preferred tools for each type of user group. The *whistle blower* prefers tools allowing anonymous submission or sharing. *Social media users* prefer Facebook, Twitter, YouTube or other channels. While *active helpers* will use social media, mobile apps, sharing platforms (any available tools), but preferably simple tools. Note that this preferred tools example does not necessarily indicate that it should be exactly this one in reality. The security and privacy features are what matters in the indicated choices of our example. Table 7 proposes five information flow models that link the user groups with predicted security requirements.

Table 7. Information flow model, privacy-security requirement

No	Information flow model	Predicted Privacy/ Security Requirement	User Group
1	Citizen → Moderator → Citizen	Anonymity or strong security and privacy	Whistle Blower
2	Citizen → No Moderator → Citizen	Medium to strong security and privacy	Social media users
3	Special Group → Moderator → Special Group	Minimum is enough	Active helper
4	Citizen-Special Group → Moderator → City	Anonymity or strong security and privacy	Whistle Blower
		Minimum is enough	Active helper
5	Citizen → No Moderator → City	Minimum is enough	Active helper

In *Model 1,* the information flows via sharing platform from citizen to citizen (**C to C**) is moderated. The intended communication of this type of users is to provide an alert about threats or dangers that if not reported, would have been unknown to other citizens. This type of information needs moderation for quality and truth validation. The platform needs to be supported by strong security and privacy. This model is likely to fit *whistle-blowers. Model 2* is unmoderated **C to C** information flow which typically intended for informing the circle of friends and family. *The social media users* belong to this second model, who are likely to be satisfied with medium security/privacy requirements. In this case, moderation is unnecessary. *Model 3* is moderated information flow from Special group to special group (**SG to SG**). The aim of the communication in this model is to voluntarily gather necessary disaster-related information as quickly as possible, and share it to other voluntary groups. The ultimate goal is to help people affected by crisis with extra useful information. To a certain degree, it may help disaster responders. Moderation in this communication model is necessary. Predicted users are "active helper" groups, who can work with minimum security or privacy. *Model 4* is the moderated information flow from citizen to city (**CSG to City**). The intended communication of this type of users is twofold. For **SG** is to inform about the resources available, critical situations that need to be tackled, or other issues that are thought necessary for the stakeholders in crisis. For **C**, the communication goal is the same as *Model 1*, i.e. to give an alert. The information flow in this *Model 4* does not need to be known by all people. The expectation is quick actions taken based on shared information. The *active helpers* and *whistle-blowers* belong to this fourth model. Thus, flexible security and privacy are highly important. In this case, moderation is necessary. *Model 5* is unmoderated Citizen to City information flow. The intended communication is to notify stakeholders their availability or their volunteer efforts in responding to disasters. This type of communication does not need moderation.

6   Conclusions, Limitations and Future Work

In this paper, we have proposed security and privacy metrics, and intuitive-based user group classifications with respect to the information and communication engagement tools. We conclude that the requirement for privacy and/or anonymity depends on the intended communication target, and this varies between the different user groups, and on the potential risk associated with a breach of privacy. The insights from the discussion in this paper is that we should mitigate reluctances of the *whistle-blower* to use any types of community engagement and information sharing. For a *whistle-blower* that sometimes carry urgent information, the risk is very high that he will be in major trouble if his privacy is violated. We also should not slow down the active helpers by making the platform too complex - e.g. through excessive security, although for *an active helper*, that risk is more like a minor annoyance. Both these groups usually want to spread the information as wide as needed to reach the proper authorities. On the other hand, *social media users* tend to target friends and family and may for example either want to tell that they are safe, or inform about local risks. This information may still be of use to the crisis handlers if it is available to the public, but reasonable privacy settings may also prevent this to happen.

Thus, the policy makers or local authority in the city should be willing to consider all relevant types of user groups in the society based on their preferred privacy and security requirements, and allow different user groups to participate through different platforms, including representative platforms from those classes of platforms mentioned in Table 6. Leaving out whistle-blowers, or slowing down and annoying active helpers would impair citizen engagement and ultimately resilience. To be able to get a complete picture from information shared by citizens, we suggest that both a specialised platform with simple verified-user messages as well as opening for moderated anonymous messages - and relevant social networks, should be utilized.

Finally, we also need to cover some limitations of our work: (1) We assume that evaluators of the security and privacy level of the engagement tools have a limited expertise on security but should know the minimum requirements to determine whether or not such criteria is fulfilled or covered. (2) The methods for evaluating security and privacy of the engagement tools are not from the insider perspective but from what information has been made available for public or is externally observable. (3) The suggested metrics are only an initial proposal. The security and privacy metrics that are relevant for city stakeholders can be elaborated further in different stages of the resilience cycle: preparedness, response, recovery and mitigation. Likewise, the matrix for the user groups can be elaborated further to include e.g. engagement tools for helping individuals that are affected by the disasters, where the security and privacy will be extremely important. For example, the engagement tools will include counselling for trauma, shocks or other psychological or psychosocial problems, or other issues that are not identified here. (4) To be aware that the strong privacy or anonymity that allows whistle-blowers to feel comfortable enough to submit their information, can also be used for actors with bad intentions, for misleading of even attempting to trap rescue personnel, or for submitting bomb threats and other criminal messages. (5) Our experiment, especially the evaluation of the availability metric is based on limited observation time, and not e.g. through monitoring over longer period, where then we could claim e.g. "uptime of 99%".

There are many directions that could be investigated further based on this study, but it should still be able to stand on its own as a set of guidelines for the security and privacy aspects of selecting tools for engaging citizens in creating a resilient society.

References

1. UN: World's population increasingly urban with more than half living in urban areas (2014). http://www.un.org/en/development/desa/news/population/world-urbanization-prospects-2014.html
2. Yang, T.-M., Maxwell, T.A.: Information-sharing in public organizations: a literature review of interpersonal, intra-organizational and inter-organizational success factors. Gov. Inf. Q. **28** (2), 164–175 (2011)
3. Palen, L., et al.: A vision for technology-mediated support for public participation & assistance in mass emergencies & disasters. In: Proceedings of the 2010 ACM-BCS Visions of Computer Science Conference. British Computer Society (2010)
4. Pipek, V., Liu, S.B., Kerne, A.: Crisis informatics and collaboration: a brief introduction. Comput. Support. Coop. Work **23**(4–6), 339–345 (2014)
5. UNISDR: Hyogo framework for action 2005–2015: building the resilience of nations and communities to disasters. In: World Conference on Disaster Reduction, Kobe (2005)
6. Trnka, J., Johansson, B.J.E.: Resilient emergency response: supporting flexibility and improvisation in collaborative command and control. In: Murray, E.J. (ed.) Crisis Response and Management and Emerging Information Systems: Critical Applications, pp. 112–138. IGI Global, Hershey (2011)
7. UNISDR: Living with risk: a global review of disaster reduction initiatives: 2004 version - volume II Annexes. In: International Strategy for Disaster Reduction (ISDR). United Nations, New York and Geneva (2004)
8. Liu, S.B.: Crisis crowdsourcing framework: designing strategic configurations of crowdsourcing for the emergency management domain. Comput. Support. Coop. Work **23**(4–6), 389–443 (2014)
9. Liza, P.: Sociotechnical uses of social web tools during disasters. In: Elayne, C. (ed.) Knowledge Development and Social Change through Technology: Emerging Studies, pp. 97–108. IGI Global, Hershey (2011)
10. Tapia, A.H., Moore, K.: Good enough is good enough: overcoming disaster response organizations' slow social media data adoption. Comput. Support. Coop. Work **23**(4–6), 483–512 (2014)

11. Lindsay, B.R.: Social media and disasters: current uses, future options, and policy considerations. In: Congressional Research Service, C.R.f. Congress, Editor (2011)

12. Avižienis, A., et al.: Basic concepts and taxonomy of dependable and secure computing. IEEE Trans. Dependable Secur. Comput. **1**(1), 11–33 (2004)

13. Liu, P., Chetal, A.: Trust-based secure information sharing between federal government agencies. J. Am. Soc. Inform. Sci. Technol. **56**(3), 283–298 (2005)

14. Schwartz, P.M., Solove, D.J.: Pii problem: privacy and a new concept of personally identifiable information. NYUL Rev. **86**, 1814 (2011)

15. Herrmann, D.S.: Complete Guide to Security and Privacy Metrics: Measuring Regulatory Compliance, Operational Resilience, and ROI, p. 848. Auerbach Publications (2007)

16. Jansen, W.: Directions in Security Metrics Research. Diane Publishing, Darby (2010)

17. Jaquith, A.: Security Metrics: Replacing Fear, Uncertainty, and Doubt. Addison-Wesley Professional, New York (2007)

18. Fernandez, E.B.: A methodology for secure software design. In: Conference on Software Engineering Research and Practice (SERP 2004) (2004)

19. Cavoukian, A.: Privacy by design: the 7 foundational principles. Implementation and mapping of fair information practices (2006). https://www.privacyassociation.org/media/presentations/11Summit/RealitiesHO1.pdf

20. Pfleeger, S.L.: Security measurement steps, missteps, and next steps. IEEE Secur. Priv. **10**(4), 5–9 (2012)

21. Ushahidi. https://www.ushahidi.com/

22. Google. https://www.google.org/crisisresponse/

23. FEMA. https://play.google.com/store/apps/details?id=gov.fema.mobile.android

24. NSW. http://www.rfs.nsw.gov.au/about-us/our-districts/mia/fire-information/fires-near-me

25. Quakewatch. https://itunes.apple.com/app/disaster-alert-pacific-disaster/id381289235?mt=8

26. EarthquakeBuddh. https://www.techinasia.com/earthquake-buddy-alert-location

27. DisasterAlert. http://www.pdc.org/solutions/tools/disaster-alert-app/

28. WikiForProfessionals. http://www.crisiscommunication.fi/wiki/Main_Page

29. Emergency2.0. http://emergency20wiki.org/wiki/index.php/Main_Page

30. PEP: Public Empowerment Policies for Crisis Management (2016). https://agoracenter.jyu.fi/projects/pep

31. Bharosa, N., Lee, J., Janssen, M.: Challenges and obstacles in sharing and coordinating information during multi-agency disaster response: propositions from field exercises. Inf. Syst. Front. **12**(1), 49–65 (2010)

32. Turner, M., et al.: Does the technology acceptance model predict actual use? A systematic literature review. Inf. Softw. Technol. **52**(5), 463–479 (2010)

33. Lee, J., et al.: Group value and intention to use — a study of multi-agency disaster management information systems for public safety. Decis. Support Syst. **50**(2), 404–414 (2011)

34. Aedo, I., et al.: End-user oriented strategies to facilitate multi-organizational adoption of emergency management information systems. Inf. Process. Manag. **46**(1), 11–21 (2010)

35. Cha, J.: Usage of video sharing websites: drivers and barriers. Telemat. Inform. **31**(1), 16–26 (2014)

36. Cottrill, C.D., "Vonu" Thakuriah, P.: Location privacy preferences: a survey-based analysis of consumer awareness, trade-off and decision-making. Transp. Res. Part C Emerg. Technol. **56**, 132–148 (2015)

37. Stolfo, S., Bellovin, S.M., Evans, D.: Measuring security. IEEE Secur. Priv. **3**, 60–65 (2011)

38. Pekárek, M., Pötzsch, S.: A comparison of privacy issues in collaborative workspaces and social networks. Identity Inf. Soc. **2**(1), 81–93 (2009)

39. Hull, G., Lipford, H.R., Latulipe, C.: Contextual gaps: privacy issues on Facebook. Ethics Inf. Technol. **13**(4), 289–302 (2011)
40. Solove, D.J.: Introduction: privacy self-management and the consent dilemma. Harv. Law Rev. **126**, 1880 (2012)
41. Twitter: Best practices for using Twitter in times of crisis (2016). https://about.twitter.com/products/alerts/helpful-assets. Accessed 15 Mar 2016
42. Wikileaks. http://www.wikileaks.com

Author Index

Printed in the United States
By Bookmasters